W9-BWQ-978

ON FERTILE GROUND

ON FERTILE GROUND

PETER T. ELLISON

HARVARD UNIVERSITY PRESS
Cambridge, Massachusetts London, England 2001

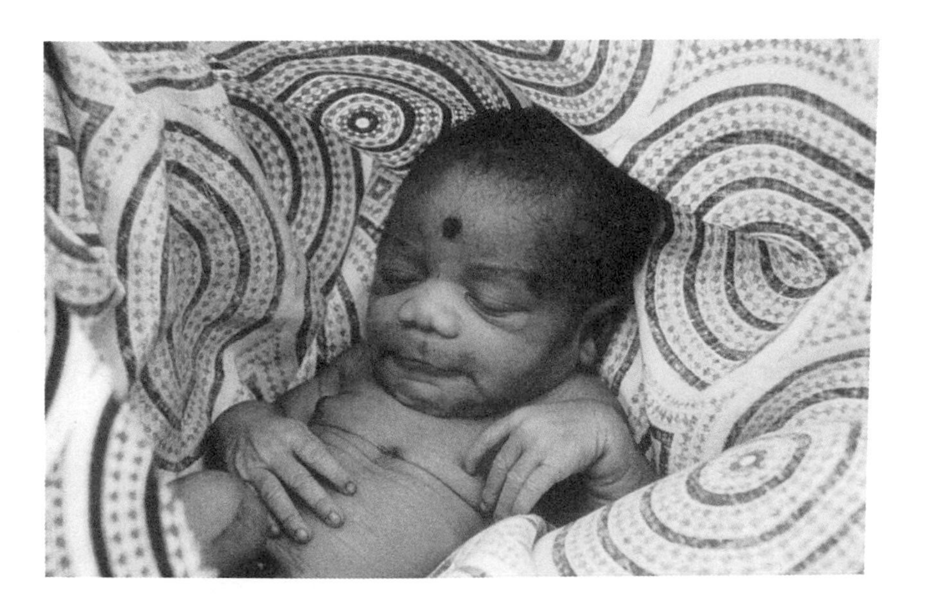

The first quotation on p. ix is from Olaf Stapledon, *Last and First Men* (London: Methuen, 1931; reprinted by Dover, New York, 1968, in *"Last and First Men" and "Starmaker": Two Science Fiction Novels*), p. 13 of the Dover edition. The second quotation on p. ix is from *Silent Spring* by Rachel Carson, p. 246. Copyright © 1962 by Rachel Carson. Copyright © renewed 1990 by Roger Christie. Reprinted by permission of Houghton Mifflin Company. All rights reserved.

Library of Congress Cataloging-in-Publication Data

Ellison, Peter Thorpe.
On fertile ground / Peter T. Ellison.
p. cm.
Includes bibliographical references and index.
ISBN 0-674-00463-9 (hardcover : alk. paper)
1. Human reproduction. 2. Human evolution. 3. Natural selection. I. Title.

QP251 .E43 2001
612.6—dc21 00-059688

For Pippi,
My partner in love, life, and reproduction

CONTENTS

When your writers romance of the future, they too easily imagine a progress toward some kind of Utopia, in which beings like themselves live in unmitigated bliss among circumstances perfectly suited to a fixed human nature. I shall not describe any such paradise. Instead I shall record huge fluctuations of joy and woe, the results of changes not only in man's environment but in his fluid nature.

—OLAF STAPLEDON, *Last and First Men*, 1931

The balance of nature is not the same today as in Pleistocene times, but it is still there: a complex, precise, and highly integrated system of relationships between living things which cannot safely be ignored any more than the law of gravity can be defied with impunity by a man perched on the edge of a cliff. The balance of nature is not a *status quo;* it is fluid, ever shifting, in a constant state of adjustment. Man, too, is part of this balance.

—RACHEL CARSON, *Silent Spring*, 1962

TWO BIRTHS

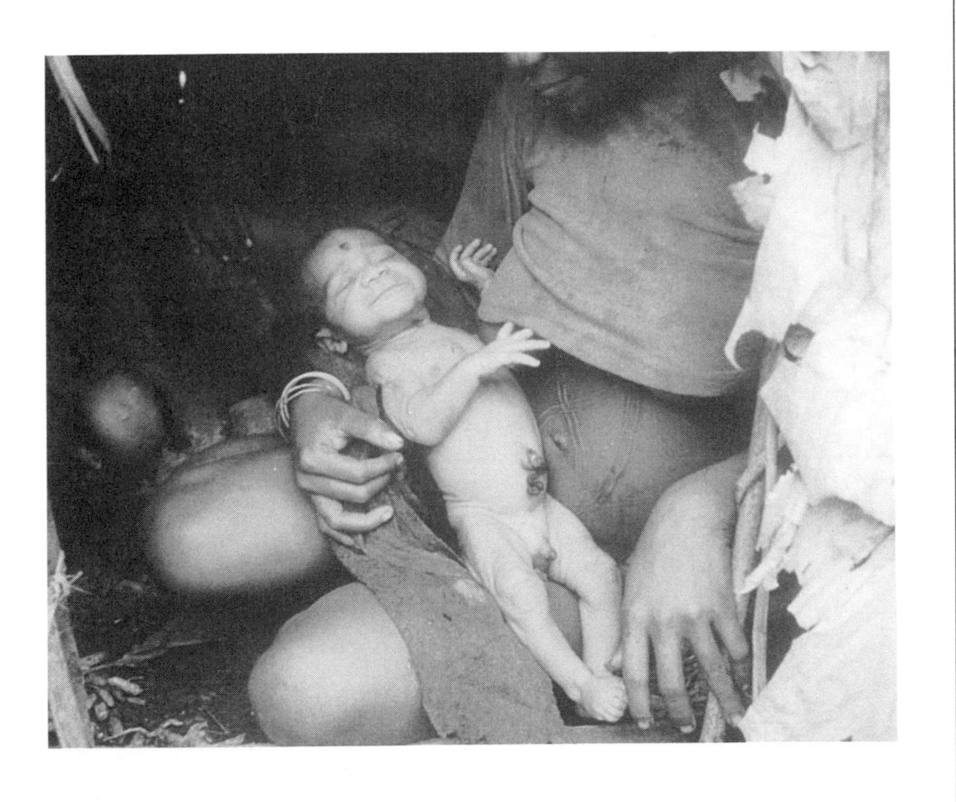

IT WAS EARLY EVENING in the summer of 1984. My wife, Pippi, and I were starting to prepare dinner. One of us stoked the fire from the stack of scavenged wood to the side of the mud oven while the other sifted insects out of the rice. We worked by the fading light filtering through the trees into the unwalled *baraza*. Twilight doesn't last long on the equator, where the sun drops vertically to the horizon. Day quickly turns into night. And night can be particularly dark in the Ituri, a dense tropical rain forest in the northeast corner of Africa's Congo basin. Except in artificial clearings made by human hands, or on the stony outcrops that jut in places above the trees, or in the middle of the broad, muddy rivers, the canopy of the forest closes overhead to hide away the moon and stars.

It was our turn to prepare dinner. The four other researchers who made up our group were in their huts transcribing field notes or slowly making their way back up from a bath in the river. Soon we would have to light the kerosene lanterns. Because of the time of day, I was surprised to see my research assistant, Kazimiri, hurrying up the path to our *baraza*. Lese villagers like Kazimiri did not like to be on the road after dusk when visibility was poor and *bilozi*, or witches, were about. But by his face, I could see that Kazimiri was worried about something worse than *bilozi*.

It was his wife, Elena. She had started her labor, even though her pregnancy was only seven months along. Nuns from the mission to the south had come through by Land Rover earlier in the day, dispensing medicine for leprosy and malaria, bouncing over the rutted track that was all that was left of the Belgian road built decades before. They had examined Elena and prodded her belly, pronouncing her carriage to be ominously low. They urged Kazimiri to bring her to the mission so that they could watch over her and tend to the birth when it came. The mission, however, was two days' walk away, and Elena was unable to make the trip. Now labor had started, and Kazimiri was desperate. Would we take them to the mission in our truck? Kazimiri had been to secondary school there. He knew the nuns had a dispensary with medicine. And he knew that it was wrong for Elena to be in labor now. He suspected that the nuns had somehow brought on the labor with their prodding fingers. But he also believed they would do what they could to help Elena and her child survive.

By this time our companions had gathered around us in the *baraza*. We conferred quickly and decided to attempt the trip. Pippi and I would go with Kazimiri and Elena and take Baudoin, another villager, along to help. We started to make preparations for the journey, sending Kazimiri back to his *parcel*, or homestead, to get Elena ready. Some of us began to siphon gas from a drum into the truck while others prepared a litter for Elena in the truck bed. Someone went for Baudoin. It wasn't a trivial trip, especially at night. It would take us five to six hours at least.

By the time we nosed the truck around the bamboo thicket and onto the scar of road the last traces of daylight were gone. The growl of the engine in its low gears carried through the night and brought Lese villagers and Efe pygmies to the side of the road to watch our departure. We couldn't always see them, but we could hear their voices calling out to Baudoin in the back of the truck and to each other. I leaned out the window to ask Baudoin what they were saying. We used Swahili to communicate since the tonal, Sudanic language of the Lese was still too difficult for us. What were they saying? Was something wrong? Baudoin tried to inject enthusiasm into his reply, but failed. No, nothing was wrong. But the birth had already happened. Twin boys.

When we pulled up to Kazimiri's *parcel* there was already a small crowd outside the hut. Kazimiri came forward to greet us and to accept our gift of clean cloth with a wan smile. Pippi took up a kerosene lantern and dashed into the hut. Soon she reemerged with sobering news. One infant was already dead and the other seemed sure to follow soon. Elena was weak and spiritless but was not bleeding or feverish. There were three female relatives of Kazimiri's with her. No men were allowed in the hut, of course, not even Kazimiri. Soon two of the women appeared in the door of the hut to display the babies. They were tiny and shriveled, frailer than any healthy newborn. One was clearly dead, but the woman holding him jiggled his limp limbs and made the tiny head rock unnaturally back and forth. The other infant breathed weakly and made tiny movements of his own. They were soon taken back inside. I glimpsed Elena sitting in a corner of the hut with her back against a post, a piece of cloth draped loosely around her body, her hands motionless in her lap, her face blank.

Outside the hut visitors spoke in low tones. No one seemed happy, but no one spoke of death either. Pippi and I wondered whether people realized that one infant was dead and asked Baudoin discreetly. They knew, he said, but they would not speak of it yet while the fate of the second boy hung in the balance. We stayed for a while, but conversation was difficult. Everyone was waiting for the second death, a death that seemed inevitable but that could not be acknowledged. Even if we could get the infant to the mission before he died there was nothing they could do for him there. The sort of emergency neonatal facilities that might help were hundreds of miles away and seemingly hundreds of years away as well. Eventually we returned to camp, leaving behind the lantern, blankets, and food. We promised to return in the morning and wished Kazimiri peace as we shook his hands in parting.

A year later I stood by Pippi's bed in an underground world of artificial twilight, stroking her forehead and listening to the hum of a fetal monitor. I knew from my watch that we had been in this birthing room for more than twenty-four hours, but the intensity of anticipation and anxiety had long since distorted my sense of time. At least the epidural anesthesia had freed Pippi from the excruciating pain of her contractions

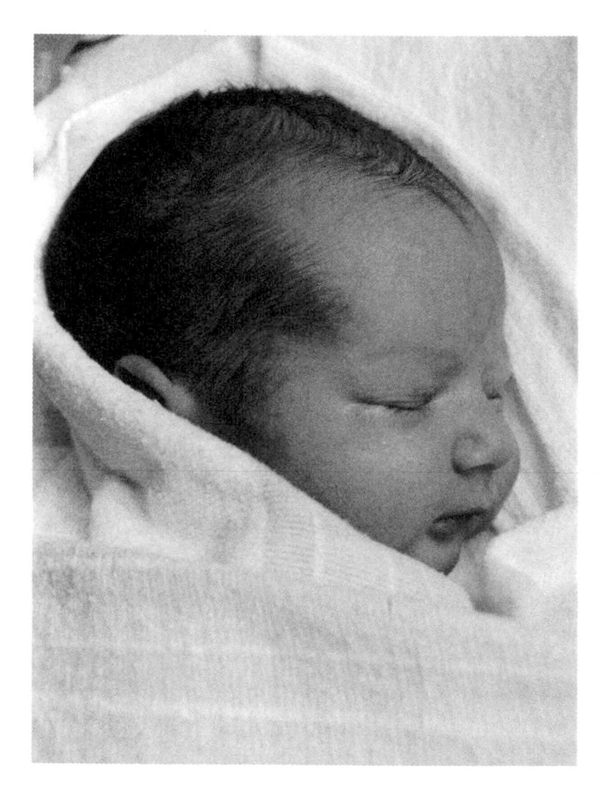

and allowed her to slip into an exhausted doze. The obstetrician had called for an operating room to try a forceps delivery. The labor was simply not progressing, the contractions were weakening, perhaps under the influence of the anesthesia, and the risk of infection was climbing. Soon the nurses arrived to transfer Pippi to the operating room. She awakened from her brief nap with a start and asked again whether I would be with her. She was reassured as she was unhooked from her drip. I was sent to change into scrubs in an adjoining room.

It took me a few minutes to find the operating room when I emerged. At one point I went back to the birthing room where we had been to try to get oriented, but it was already occupied by another woman and her anxious partner. Finally, a nurse came and led me from the twilight into the sudden glare of the operating room. Pippi was on the operating table, a curtain draped across her middle. She reached for me and took my hand. I should have noted the time on the large clock on the wall, but didn't. So I don't really know how long the obstetrician tried to deliver the baby. Finally, though, he said he would have to do a Caesarean section. The baby's head simply could not pass through the birth canal. Pippi had already been prepped and the anesthesia had already taken effect. The doctor and his assistants worked efficiently behind their curtain, and suddenly, before we realized the surgery had even begun, he was with us: our son, Sam: bloody, ruddy, with a head still compressed by the effort made on all sides to force him through his mother's pelvis. Like all new parents, our lives, too, began again at that moment.

NATURAL SELECTION AND REPRODUCTIVE PHYSIOLOGY

How different can two birth stories be? One seems to represent the miracles of a modern age, the other the harsh realities of a primeval world. If Elena had been transported half a world away to a hospital in Boston her twin sons might easily have lived, might be strapping teenagers now, a class ahead of Sam in school. If Pippi had been transported half a world away to a mud-floored hut in the Ituri Forest, she almost certainly would have died, and Sam would never have been born. But both of these stories point to another story, the story of the shaping of human procreation. Both point to the hazards inherent in reproduction itself, the risk of death that shadows new life. Both point to constraints that must be accommodated: the physical difficulty of passing a full-term head through a female pelvis, the difficulty of growing two fetuses to full term. Both whisper of the pitiless process of natural selection that has shaped all features of our biology, reproduction included. We are the descendants of those who survived, those who succeeded in giving birth. Elena lost not only two infant boys, but two grown sons that might have been, the grandchildren they might have fathered, the lineages they might have founded. All the copies of genes the twins carried and might have passed to their own offspring were buried with them, among them perhaps genes that increased the probability of twinning. Because of the intervention of a C-section Sam's genes live on, perhaps among them genes for a large fetal head size or a narrow maternal pelvis. Dramas like this, repeated countless billions of times in our evolutionary history, have endowed us with our physiological capacities and limitations and shape us still.

This book is about the story behind the stories of Elena's twins and Sam's birth. It is an attempt to understand not only how human reproduction works but why, to see in the details of our reproductive physiology the imprint of natural selection and the trace of our evolution. Reproduction is a central feature of life, perhaps its most distinguishing characteristic. It is also the force that drives biological evolution. Darwin envisioned a biological world in which species of organisms exist not as perfect archetypes but as a blur of individual variation, each of those variants competing (however unconsciously) to perpetuate itself in the next generation through reproduction. The variants that have the greatest reproductive success come to predominate. Those with the least reproductive success disappear. Through time the species evolves, shaped by the differential reproductive success of its members.

Any feature of biology that can influence the reproductive success of individuals can be shaped by natural selection. The impact can be very indirect and faint. Perhaps having eyebrows helped our ancestors keep sweat out of their eyes and thus perform a variety of functions better that required good vision while working, some of which may have helped them to survive a bit better than those without that trait. Perhaps eyebrows made a difference in only one in a hundred lives. But that tiny advantage, multiplied over the billions of people who have lived on the earth, may have helped the trait to spread. For other traits with a more immediate connection to survival and reproduction the impact can be great and the selection drastic. Malformations of vital organs, missing critical enzymes, disabled immune systems: these are traits that would ordinarily lead to rapid death in the absence of medical intervention. Selection on the reproductive system must be particularly strong, since variation in reproductive function often translates directly into variation in reproductive success. Reproductive physiology, therefore, provides particularly fertile ground for the study of evolution.

But the study of evolution always necessarily involves the study of ecology. In the analogy of the naturalist Marston Bates, if evolution is the play, ecology is the stage. Evolution always unfolds in a particular ecological context. Adaptation between an organism and its environment, Darwin argued, is the outcome of natural selection because the functions that an organism must perform to achieve reproductive success are always environment specific. A gibbon's limbs are wonderfully suited to locomotion in the forest canopy, a zebra's to locomotion over the savanna, and a seal's to locomotion in water. Rotate each of these animals into one of the others' habitats and their elegantly designed limbs become awkward encumbrances. Making inferences about natural selection thus always involves making inferences about the environment in which the selection occurred. And since the features of contemporary organisms are the result of an evolutionary history stretching into the distant past, this often involves making inferences about environments that have long since disappeared or been transformed. Nevertheless, if we are to understand human reproductive physiology as a product of evolution, we must understand it as well in relation to the ecological contexts of human beings, past and present. Broadening our perspective to appreciate the range of environmental circumstances under which contemporary human beings live and reproduce can help. Too often human physiology is taught and studied under a relatively narrow range of conditions typical of urbanized society in industrialized countries. Understanding how human physiology operates, even today, in other ecological contexts can help us project our understanding into the environments of our formative past.

We must also appreciate at the outset that physiology by its nature involves the ability to adjust and respond to changing environments. Physiological processes, especially those of crucial importance to survival and reproduction, are rarely if ever fixed and unchanging. Even as the sort of limbs that are good for locomotion in one environment are not necessarily good in another, so the responses of the human reproductive system that are optimal in one environment are not necessarily optimal in another. When food is plentiful and likely to remain so, the extra physiological burden of pregnancy may be relatively easy to bear. When food is scarce or unpredictable, it may be more prudent to conserve resources and avoid pregnancy. Over our evolutionary history the range of environmental variation, its pattern, and its predictability have selected for physiological mechanisms to track that variation and continuously adjust to it.

But not only does the environment change, and with it the physiological responses of the organism, the organism changes as well over the course of its life. Some of that change is accidental or idiosyncratic, unique to the individual. Other change is predictable and common to the lives of most or all members of the species. Among the most predictable features of constitutional change—change in the individual rather than change in the environment—is the process of aging. Age is a principal vector for all human biology. It is the axis along which our development unfolds and the inexorable measure of all our opportunities, including our opportunities to reproduce. It also weights and colors those opportunities differently in terms of their costs and benefits. The risks we are willing to take, the sacrifices we are willing to make, and the rewards for which we are willing to make them change at different points in our lives, whether we are talking about conscious decisions regarding career alternatives or physiological decisions regarding metabolic investments in reproduction. Natural selection has shaped our physiology to adjust to our changing age as well as our changing environments. Because age unfolds in such a predictable course in all humans, age-related changes in physiology are particularly predictable as well. But that does not mean that they are narrowly constraining. All fifty-year-olds are not the same, and the average fifty-year-old is not the same in different environments. Nevertheless, the average fifty-year-old in every environment is closer to death and more likely to have grown children than the average twenty-year-old. These and other inescapable correlates of age, of what an evolutionary biologist refers to as life history, have helped to sculpt our physiology throughout our evolutionary past.

MOLECULAR MESSENGERS

The regulation of reproduction involves some of the body's most exquisite machinery, ranging from special anatomical features to the hidden secrets of our genes. Prominent in the drama of reproductive physiology are hormones. Hormones are molecules that carry information from one place in the body to another, signaling the cells that receive the information to change their biochemical activity in specific ways. The system of hormonal transmitters and receivers, the endocrine system, is not the only system of information transfer in the body. The central nervous system is probably the most familiar of our physiological information networks. The immune system is another. All three of these systems have certain features in common, features that are extremely ancient in the history of life on this planet. They all take advantage of the ability of protein molecules to assume a virtually infinite variety of physical shapes, some of which provide binding sites, places on the molecule where other molecules can fit quite precisely. The molecule providing such a site is referred to as a receptor and the molecule that fits as a ligand. An analogy is often drawn between the receptor and its ligand on the one hand and a lock and its key on the other, stressing the close complementarity of the two shapes and the specificity of the pairing. Like a key turning in a lock, the binding of a ligand to its receptor can cause a temporary change in the receptor molecule, a change that can alter its shape or chemical characteristics elsewhere on its length than at the binding site itself. That change in turn can trigger biochemical events involving other molecules that have remained latent in the absence of the ligand. The tumblers inside the lock turn: the door opens.

The central nervous system uses this mechanism of ligand and receptor to pass information from one nerve cell, or neuron, to the next across the synapse separating them. An electrical impulse arriving at the end of one neuron causes it to release from its tip molecules known as neurotransmitters that travel across a microscopic space to bind to receptors on the receiving neuron. The binding of the neurotransmitters triggers biochemical events that lead to the transmission of an electrical impulse down the length of that neuron to the next synapse. Eventually, at the end of the neuronal pathway, the release of neurotransmitters can cause a muscle fiber to twitch or feed information to the processing centers of the brain. The central nervous system is the major system regulating our behavior, allowing us to react to our environment both reflexively and reflectively. It is our most rapid regulatory system—the hand springs back from the hot stove before the pain even registers—and our most flexible. It allows us to respond to a nearly limitless variety of situations and contingencies. As a system of physiological regulation, however, it is characterized by location specificity of information transfer. Signals travel from point to point over a network of physical connections, the network of neurons that "wire up" our bodies. The location specificity of the information transfer is integral to the operation of the central nervous system, of course. Efficient locomotion depends on a smooth sequence of contractions in specific muscles, not general convulsions. Efficient sensory perception depends on topographically specific stimulation of sensory neurons, lest we pull away the wrong hand from the fire. The central nervous system is like an old-fashioned telegraph, relying on a network of physical wires to carry messages from place to place. The messages travel very quickly and with high geographic specificity, but they can't go where the wires haven't been strung.

In contrast, the endocrine system is more like a radio or television broadcast system. Signals are released from a point of transmission in a geographically nonspecific manner. They travel everywhere. But even though the electromagnetic waves that carry the information from all those radio and TV stations are passing around and through us all the time, we have no awareness of them, *unless* we have a receiver tuned in to receive the signal. In a similar fashion, endocrine glands release specific molecules—hormones—into our general circulation. They are carried throughout the body by the circulatory system, potentially reaching a large majority of our cells. But not all cells respond; only those respond that have receptors "tuned" to the signal carried by the particular hormonal ligand. This system is especially effective in communicating a given signal to lots of target cells, which may be located in many different places. The responses of those targets can be quite different as well, but they will be coordinated by their relationship to the common signal. Many signals can be passing through the body simultaneously. Just as each radio picks up only the frequency to which it is tuned, each cell responds only to the hormones for which it has receptors. The endocrine system is the system that the body uses to achieve integrated responses among various cells, tissues, and systems, integration that is necessary to many critical biological processes, such as growth, metabolism, and reproduction. If the central nervous system is the principal regulator of our behavior, the endocrine system is the principal regulator of our physiology.

Hormones will necessarily surface often in the course of this book. With their complex-sounding names they carry an aura of biochemical complexity and can seem beyond the grasp of the uninitiated. This needn't be the case. Those hormones that are important to the story of human reproduction will be introduced as needed and in ways that will make the molecules themselves as well as their functions understandable in simple terms. At the outset, however, it may be useful to note that we will principally be dealing with two major classes of hormones: proteins and steroids. Protein hormones are the most common and include the hormones secreted by the pituitary gland at the base of the brain, such as the gonadotropins that help to regulate the ovaries and testes. Proteins are large molecules composed of chains of amino acids strung end to end and folded up in complicated conformations. They are soluble in water and hence in blood, but not in lipids, the constituents of fats. Because of their size and their lipid insolubility they cannot pass through cell membranes. Cells that respond to protein hormones, therefore, display the receptors for them on the surface of their membranes. When those receptors interact with one of their ligands, changes in the receptor trigger changes inside the cell, usually involving the activation of enzymes, biochemical machines that carry out the cell's business.

In contrast to proteins, steroid molecules are tiny. Steroids are produced only by the gonads—ovaries and testes—and the adrenal gland, and include the so-called sex hormones, the androgens, estrogens, and progestins. They are composed of eighteen to twenty-one carbon atoms bound together with associated hydrogens and oxygens and have only a small fraction of the mass of the much larger protein hormones. They are soluble both in water and in lipids, being very lipid-like themselves. As a consequence they can travel through cell membranes as if they weren't there, entering into the interior of a cell and even into its nucleus. The cells that respond to steroids tend to produce receptor molecules that stay in the interior of the cell. When they interact with their steroid ligands, the receptors often change so as to bind with the genetic material in the nucleus of the cell to affect the process of gene expression.

Steroid hormones also travel to areas of the body that are off limits to protein hormones. They can cross the blood-brain barrier, for instance, carrying information from the periphery of the body to the central nervous system. But for the same reason steroid hormones by themselves do not stay in the "pipes" provided by the circulatory system. Rather than being carried by the blood, a steroid molecule has a natural tendency simply to diffuse outward from its site of production to neighboring cells and tissues. Special "carrier proteins" are needed to bind to steroid hormones so that they can be carried efficiently in the blood to target tissues that may be quite distant. Because there is a chemical equilibrium at every point along the way between the "protein-bound" steroid and the "unbound" or "free" steroid, all tissues are exposed to essentially the same concentration of free steroid, rather than being exposed to a diffusion gradient of concentration that declines with increasing distance from the site of production. A final consequence of their lipid solubility is that steroid hormones can be measured in saliva as well as in blood. Only free steroid hormones enter the saliva, but that is the same fraction that enters a cell to interact with steroid receptors. The ability to measure these potent physiological regulators in a medium as easy and painless to sample as saliva has provided important opportunities for expanding our understanding of reproductive physiology beyond the confines of the hospital and the clinic.

Hormones are perhaps the minutest elements of the human reproductive system. But the events they regulate—the turning on and off of specific genes, the speeding up or slowing down of specific chemical reactions—have effects that cascade upward through our biology to result in some of the most profound features and moments of our lives. These are features and moments that define and transform us as individuals, casting us in our most important roles, as men and women, mothers and fathers, offspring and siblings. They generate the fabric of kinship from which our communities and societies are woven. These same processes drive our evolution from its primeval past toward its unseen future. By understanding the ecology of human reproduction, we can understand ourselves more fully and can perhaps catch glimpses of the path behind us and the horizon in front.

SURVIVING THE FIRST CUT

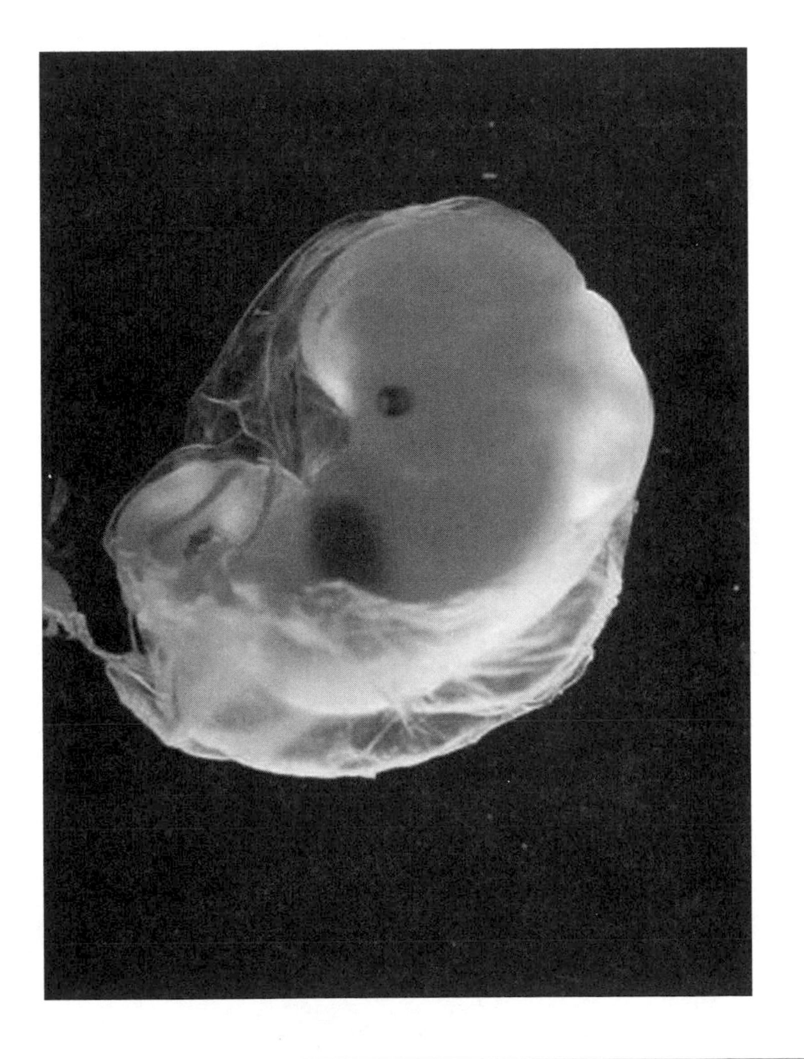

HIGH IN THE FALLOPIAN TUBE of the mother, a new human genome has been formed. A single sperm from the father has managed to find the egg cell and, with the help of a special set of enzymes that it kept under its "cap," to dissolve its way through the veils of gelatin and protein surrounding the egg, to fuse with the egg's cell membrane, and to inject the contents of its nucleus into the egg. The egg cell on its part has sealed off its cell membrane to prevent the intrusion of other sperm cells, has ejected half of its freshly replicated chromosomes, and has paired the rest with those delivered by the successful sperm. This cell, no longer an egg but an embryo, is now genetically unique. From two human beings has been formed the beginning of a third, completely new, an individual that has never before existed and never will again. The drama of its life stretches away like a path with infinite twists and turns. But an immense amount of work must be done first to create a human being from this single cell. As if summoning its energy and resolve for the task ahead, the cell continues its slow creep down the fallopian tube toward the uterus without any visible change. For nearly twenty-four hours nothing happens. Then the first cell division occurs.

The next few days are among the most dangerous that this new organism will ever face. It has several crucial tasks that it must perform successfully or its brief existence will come to an end. It must send a signal to its mother's reproductive system that will override the sequence of events normally leading to menstruation. It must establish a connection with its mother's blood supply in order to obtain food and oxygen. And it must evade its mother's immune system in the meantime.

LYING LOW

The last of these tasks must be taken up first. As soon as its maternal and paternal chromosomes are joined, the new organism is genetically distinct from its mother and vulnerable to attack by her immune system. An army of lymphocytes, specialized white blood cells, patrols the mother's body searching for the presence of foreign organisms. When found, they are targeted and destroyed with ruthless efficiency. Vigilance is particularly high where invasion is most likely: in the mouth, throat, and gastrointestinal tract; in the nasal passages, trachea, and lungs; and in the reproductive tract. Unlike the male reproductive tract, the female tract is not a blind alley ending in the gonads, but an open passage to the woman's body cavity. Ordinarily this passage is particularly well protected. Mucus secretions guard the opening of the cervix, trapping any pathogen seeking to pass from the vagina into the uterus. An acidic environment is maintained within the vagina that is hostile to most pathological bacteria. Lymphocytes patrol the passage, clustering at the cervix and the openings of the fallopian tubes into the uterus.

The sperm cell that succeeded in fertilizing the egg had itself to overcome these barriers, aided by the mother's own physiology. The events leading to the release of the egg cell from the mother's ovary also caused changes in her reproductive tract that helped the sperm pass through it. The mucus secretions from the cervix, ordinarily producing a thick, viscous mesh of interlacing strands of mucin, became thin and watery. The mucin strands untangled and, rather than forming a dense and impenetrable mat, lay more or less parallel to the long axis of the cervix, allowing sperm cells to enter the uterus. The acid environment of the vagina was rapidly neutralized by the carbonates contained in the father's semen. The mother's immune response was temporarily suppressed. A brief window of opportunity opened for the sperm to pass before the barriers went up once again.

But now the developing embryo is also a potential target for the mother's immune system. As the embryo begins to decode its own genetic blueprint, it reveals its identity as a genetically distinct organism. The surface of its cells becomes speckled with protein fragments and specialized markers that will allow its own cells to organize and interact appropriately with each other. But because these markers are transcribed from the embryo's unique set of genes, they potentially identify it as foreign to its mother's immune cells. It tries, with the mother's help, to hide these markers. Initially the egg cell and the subsequent embryo are surrounded by a layer of the mother's own cells, cells that came from the ovarian follicle in which the egg cell matured and that escaped from the ovary along with the egg at the time of ovulation. These cells are known as the cumulus oöphorus, or "egg-bearing cloud," from their appearance under the microscope. They hide the embryo from the mother's immune system for a while. Inside the cumulus oöphorus is a gelatinous layer, the zona pellucida, which also helps to obscure the protein markers that would identify the embryo as "other." But both of these layers must soon be shed in order for the embryo to efficiently absorb nutrients across its surface and ultimately to implant itself in the mother's uterus and gain access to her blood supply. As these coverings come off, the vulnerability of the embryo increases.

But why should the mother seek to eliminate her own offspring? Can't she tell the difference between a pathogen and her own progeny? No, she cannot. How could she? In the development of her own immune system she has produced an army of cells that can recognize her own cell markers. This was accomplished by a lengthy process of trial and error through which potential lymphocytes that proved unable to recognize the mother's own cell markers were eliminated. But there is no process by which the unique genetic attributes of offspring yet to be conceived could have been anticipated.

Although the mother's immune system cannot be preprogrammed to recognize its future progeny, her physiology can respond to the likelihood of an embryo's presence. The vigilance of the immune system is relaxed slightly in the week or so following ovulation under the influence of the hormones that are preparing the woman's body for a potential pregnancy. This response represents a compromise between opposing selective forces. A woman who does not suppress her active immunity at all might have a hard time getting pregnant and so leave fewer offspring with that trait to succeed her than her contemporaries who do undergo immune suppression. On the other hand, a woman who suppresses her immunity too much may run too great a risk of infection and death. This risk may seem minimal to those of us who have grown up in a world of antibiotics, frozen food, and purified water, but in the world of our ancestors, as in many parts of the world today, the risk of infection was always a serious and potentially life-threatening matter.

Mild immune suppression occurs following ovulation in every menstrual cycle, but not every menstrual cycle leads to pregnancy. When there is no embryo in a woman's reproductive tract, this immune suppression represents a costly risk with no potential benefit. Presumably it occurs because the woman's body cannot, in the first days following ovulation, be sure whether an embryo exists or not. Mild immune suppression represents a cautious reaction to the uncertainty of the situation. But the embryo can try to further suppress the mother's immune system. Among the very first excretory products that the embryo manufactures are molecules known as pregnancy-related proteins, some of which appear to act as local immune suppressants, inhibiting the interaction of lymphocytes and their targets in petri dish experiments. One of these is known as early pregnancy factor, minute traces of which can be detected in the mother's blood within forty-eight hours of fertilization. The embryo enters the cavity of the mother's uterus as a hollow ball of cells some five days old known as a blastocyst and begins to burrow into the uterine lining. By then these substances may be helping to keep the embryo from being rejected like a foreign tissue graft.

SYNCHRONIZE WATCHES

The process by which the embryo becomes imbedded in the inner lining, or endometrium, of the uterus is known as implantation. Successful implantation, not the mere fertilization of the egg, is the true initiation of a potentially viable pregnancy. Many would-be embryos may pass through a woman's uterus in her lifetime without succeeding at this task, either because the embryo itself is defective or because the endometrium is at an inappropriate stage of its own developmental cycle or is pathologically disabled. Successful implantation depends on the active interaction of the embryo and the mother. It is not something that one does to the other. The mother's endometrium must be receptive to the embryo, and the embryo must be able to implant itself. Of course outside interference can disrupt this process: an intrauterine device (IUD), for example, exerts its contraceptive effect by preventing the endometrium from responding appropriately to the presence of an embryo even though that embryo might be viable.

Implantation of a human embryo involves a deeper penetration of maternal tissues than in most other mammals. In physiological terms, the implantation is more invasive, and will eventually result in embryonic tissue bathing directly in pools of maternal blood rather than merely snuggling up close to maternal blood vessels. The endometrium has prepared itself for this event over the preceding several weeks. The uterine lining that developed a month previously had presumably not received a viable embryo and was sloughed off in menstruation. As soon as the previous lining was shed a new endometrium began to grow, stimulated by the estrogen hormones secreted by the ovary. These hormones are produced by the follicle, or cluster of cells, in which a new egg cell, or oocyte, is maturing in preparation for ovulation. The estrogen hormones that the follicle produces direct both the development of the egg and the development of the endometrium that will receive any embryo that may be produced from that egg. In this way the two processes are synchronized. The estrogen hormones from the ovary stimulate the process of cell division, or mitosis, in the endometrium so that the volume of tissue increases dramatically. Blood vessels branch out through the developing tissue, and groups of cells differentiate and form secretory glands that empty into the cavity of the uterus.

As the uterus develops, so do the follicle and the egg cell it contains. The bigger the follicle the more estrogen it secretes. The more estrogen it secretes, the faster it grows, since the estrogen also stimulates mitosis in the follicle's own cells. Inside the follicle the concentration of estrogen is particularly high, stimulating the nucleus of the egg cell to continue the complex process of replicating its own chromosomes, choosing one of each pair, and preparing the machinery that will be necessary for combining those chromosomes with a complementary set from a sperm cell. This process of nuclear maturation must be successfully completed in order for the egg cell to be fertilizable.

But in addition to directing the growth of the follicle, the maturation of the egg, and the development of the endometrium, the estrogen hormones of the ovary travel through the woman's circulatory system and interact with receptors elsewhere in her body, including some in her brain. The concentration of receptors for these hormones is particularly dense in the area at the base of the brain called the hypothalamus and an associated bit of glandular tissue called the pituitary gland that hangs off the base of the brain like a tiny holiday ornament. The pituitary is in fact a hybrid structure formed from two different tissues that become squeezed together during embryonic development within the confines of a tiny cavity at the base of the skull. A dissected pituitary can easily be teased apart into its two halves, anterior and posterior. The anterior pituitary originated from a pouch of tissue that developed in the roof of the esophagus of the embryo and migrated up to its present position. The posterior pituitary developed as a tiny finger of hypothalamic tissue that pokes down from the base of the brain and into the anterior pituitary like a finger into a soft balloon. The posterior pituitary, reflecting its origin, maintains a direct connection with the hypothalamus via nerve cells and their wirelike projections. The anterior pituitary, reflecting its origin, has no such direct connection with the hypothalamus. It does have a special route of communication with the brain, however, provided by a tiny system of blood vessels only a centimeter or so long. These vessels extend from the hypothalamus into the body of the pituitary with a finely divided bed of capillaries at both ends. This tiny bit of vasculature is known as the hypophyseal portal system (after an alternative name for the pituitary, the hypophysis). Thus although electronic signals cannot be flashed from the brain to the anterior pituitary as they can to the posterior pituitary, brain cells can secrete special hormones into the hypophyseal portal system to interact with receptors at the pituitary end. Because of the small distances involved, these signals can be sent and received very quickly. But the signals are rapidly diluted and then lost when they pass into the general circulation.

All kinds of messages pass between the hypothalamus and both parts of the pituitary. The pituitary serves as a gateway between the central nervous system and the hormonal system, with the pituitary's own hormonal secretions involved in the regulation of many of the body's crucial physiological functions. One of these involves the regulation of the gonads, both ovaries and testes. The anterior pituitary secretes a pair of large protein hormones called gonodotropins, because of their function in stimulating gonadal activity. In order for these two hormones to be produced in appropriate amounts, a particular signal must be passed from the hypothalamus through the hypophyseal portal system. This signal comes in the form of a small peptide molecule called gonadotropin-releasing hormone, or GnRH. In an adult, GnRH is normally released in a pulsatile pattern, once every hour or so. As long as it is, the pituitary responds with the production and release of its two gonodotropins, follicle-stimulating hormone and luteinizing hormone respectively, or FSH and LH for short. FSH stimulates the development of follicles in the ovary while LH stimulates the production of steroid hormones by those follicles. In addition to steroids, the follicles also produce a protein hormone called inhibin. Inhibin helps to prevent the development of supernumerary follicles by suppressing the production of FSH after one follicle has emerged as the primary candidate for ovulation.

Since steroid hormones pass through cell membranes with ease, they are ideal messengers from the periphery across the blood-brain barrier to the central nervous system. Estrogen hormones, the most potent of which is called estradiol, flow into the brain in increasing amounts as the developing follicle grows. In many animals these increasing levels of estrogens can produce dramatic changes in behavior related to reproduction. Many female mammals go into "heat," or "estrus," actively soliciting copulations from males. Activity levels may increase, with caged female rodents increasing the amount of time spent running in their exercise wheels and wild rodents extending their daily ranges and increasing the probability of encounter with a male. Even in humans subtle changes have been documented in cognitive abilities, olfactory and taste acuity, and perhaps even sexual motivation. The secretion of testosterone by the male testes affects male behavior as well, as we shall see later. Steroid hormones thus serve as important vehicles for coordinating the reproductive behavior of an organism with its reproductive state.

The estrogen levels produced by the ovary also affect the activity of the hypothalamus and pituitary and the resulting secretion of gonadotropins. As the levels of estrogen circulating in the woman's body rise, the amount of gonadotropins released diminishes, an effect referred to as negative feedback. The more estrogen the pituitary is exposed to, however, the more is required to continue to hold gonadotropin secretion in check. As long as estrogen levels are rising, this restraint is maintained. But as the follicle reaches its maximum preovulatory volume its growth slackens and estrogen levels plateau. At this point the dam that has been holding back gonadotropin secretion appears to burst and a great surge of both FSH and LH courses through the woman's body from her pituitary. The LH surge triggers digestion of the membrane surrounding the developing follicle and release of the egg it contains in the event known as ovulation. It also causes the follicular cells left behind to undergo dramatic changes. They rapidly become vascularized and accumulate stores of cholesterol, causing the whole mass of cells to appear yellow under the microscope. The cholesterol is used as raw material for what now becomes the most active endocrine gland, pound for pound, in the human body: the corpus luteum ("yellow body"). These erstwhile follicular cells continue, after a brief interruption, to produce tiny amounts of estrogen. But their major product is a different steroid hormone: progesterone. As its name implies, progester-

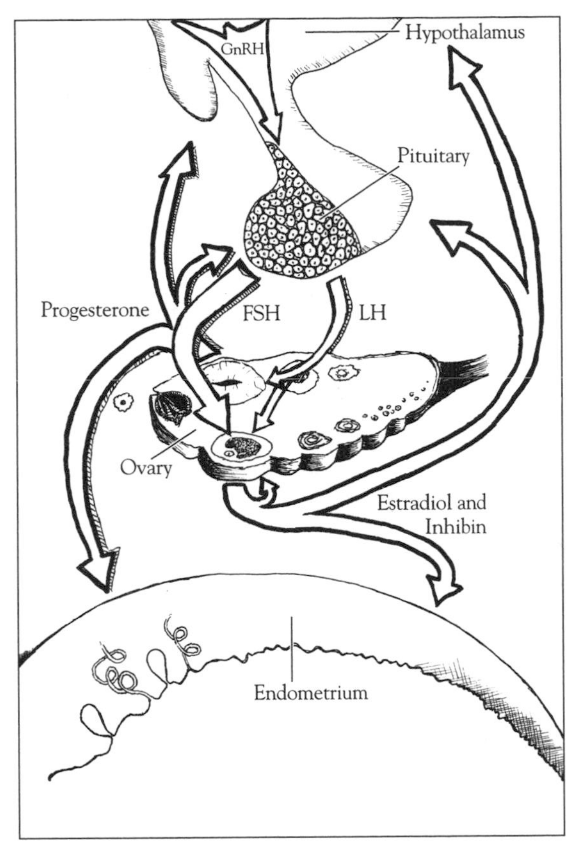

Major components of the female reproductive axis. GnRH is released in pulses from the hypothalamus, stimulating the production and release of FSH and LH from the anterior pituitary. Together FSH and LH stimulate the development of follicles in the ovay, leading to ovulation and the formation of a corpus luteum. The developing follicle produces estradiol, which supports further follicular growth, promotes the development of the endometrium in the uterus, and modulates hypothalamic and pituitary activity. The developing follicle also secretes inhibin, which suppresses the release of FSH, preventing the development of supernumerary follicles. After ovulation the corpus luteum produces progesterone, which maintains the endometrium and promotes its secretory activity in preparation for the implantation of an embryo. Luteal progesterone also modulates hypothalamic and pituitary activity.

one has a major role to play in promoting any gestation that may result from the newly ovulated egg, and it is produced in concentrations a thousand-fold greater than the estrogens produced in the preceding weeks.

The major target for the progesterone streaming out of the corpus luteum is the endometrium of the uterus. It arrests the mitotic divisions that have been bulking up the endometrium and causes the cells to undergo final differentiation and to begin the functions necessary for supporting and receiving an embryo. The secretory glands in the endometrium begin to exude nutrients that the embryo will require until it can achieve a connection to the mother's circulation. The blood vessels begin to form bulbous sinuses, and the upper layers of the endometrium set up the biochemical triggers that will allow them to respond to an embryo with the appropriate sequence of responses. In initiating all these changes, collectively called decidualization, the endometrium commits itself to its fate. Its cells have become terminally differentiated and cannot return to the mitotically active state of an undifferentiated cell. If an embryo does not implant in the next five to seven days, the endometrium will have to be shed in menstruation and a new endometrium will need to be prepared.

The coordination of the two events, ovulation and decidualization of the endometrium, is crucial. The egg and the endometrium are each launched on a course that will unfold along a narrowly fixed timetable. Their paths must cross when each is exactly ready for implantation or the enterprise will fail. We have known that the timing is crucial in experimental animals for some time. In mice, for example, implantation will only have a high chance of success if the embryo and endometrium are each of exactly the right age. If you try to implant an egg of the right age in an endometrium even a day too old or too young, the probability of success drops to nearly zero. Likewise, if you try to implant in an endometrium of the right age an embryo that is either a day too old or a day too young, the probability of success drops to nearly zero.

More recently we have learned that a similarly narrow window of opportunity exists for human implantation as well, the information coming from the practice of in vitro fertilization. There appears to be a window of only two days during which implantation has much of a chance of succeeding. An egg that is fertilized late, or that is transported slowly to the uterus down the fallopian tube, will miss this opportunity. So will an embryo that makes the journey on time, but that has developed more slowly than it should. Similarly, an endometrium that develops too quickly or too slowly will be unlikely to support successful implantation. These constraints can, in fact, make the proper timing for introduction of the embryo in in vitro fertilization quite tricky. The same techniques that are used to hyperstimulate the ovary to produce more than one egg can also hyperstimulate the endometrium.

THE BIG DIG

If all has gone according to plan—if the egg was fertilized soon after ovulation and proceeded to develop on the proper schedule, if the decidualization of the endometrium was launched in synchrony with ovulation and has proceeded apace, supported by progesterone from the corpus luteum—the embryo that begins to settle on the surface of the uterus and the uterus itself are both ready to begin the delicate dance of implantation. A series of chemical signals passes back and forth between the embryo and the endometrium, and if either fails to do its part, the enterprise can founder. The uterus secretes enzymes that help to dissolve away the zona pellucida from around the embryo. As the zona is shed and oxygen and nutrients flow into the embryo in increasing amounts, the embryo's metabolic activity increases, releasing carbon dioxide. Increased levels of carbon dioxide initiate a "suicide" sequence in the surface cells of the endometrium, causing them to undergo programmed cell death, or apoptosis.

As the endometrial cells die, cells from a portion of the embryo called the trophoblast begin to proliferate and grow downward into the endometrium. In its early stages the trophoblast can't even be said to be undergoing classical cell division, because it does not develop traditional cell membranes. It grows instead as a mass of protoplasm with many nuclei. Perhaps the absence of cell membranes facilitates the rapid diffusion of nutrients and gases through its protoplasm. For at this stage in its development the embryo is in desperate need of both nutrients and oxygen. Initially, as a newly fertilized egg, it had a certain amount of nutrients "on board" in its own cytoplasm, but these became rapidly depleted as it developed. New nutrients had to make their way into the embryo by diffusion from the surrounding body fluids of the mother. Diffusion of large molecules across cell membranes is difficult and must often involve active transport mechanisms that themselves require energy. As the number of membranes that must be crossed to reach the inmost cells increases, the efficiency of this process declines markedly. Early on, the developing embryo has to become a hollow ball of tissue, rather than a solid mass, to enable all its cells to receive adequate nourishment. But as implantation begins, the embryo is approaching the point at which more complex embryonic processes must begin, starting with the differentiation of multiple tissue layers, and simple diffusion of nutrients and gases will no longer suffice, even from the rich broth that the endometrium secretes.

Implantation in many mammalian species is a fairly superficial affair. The embryo lays itself down on the surface of the uterus and the uterus surrounds the embryo with blood vessels. As the embryo develops it grows blood vessels of its own, which become closely enwrapped by the maternal vessels. Nutrients and gases pass from the maternal circulation into the fetal circulation along these blood vessels much the way waste products pass from the blood vessels of the kidney into the renal tubules they surround. Many of the most important molecules—simple sugars like glucose and gases like oxygen and carbon dioxide—can pass between mother and fetus by simple diffusion, relying on the presence of a concentration gradient to help them flow from where they are more abundant to where they are less plentiful (though this diffusion may often be aided by active transport mechanisms as well). Often maternal and fetal blood flow past each other in opposite directions, producing what is known as a counter-current diffusion system. By matching the end of the maternal circulation with the highest concentration of nutrients (say, 100) with the end of the fetal circulation that also has the highest concentration (say, 98), and the lowest (5) with the lowest (3), this system enables transfer of nutrients to occur along the entire length of the vascular network, maximizing the percentage of nutrients that can actually be transferred (95 out of 100 in this example). If the circulation of mother and fetus were to flow in the same direction (incoming fetal blood at 2 matched with incoming maternal blood at 100) nutrient transfer would cease somewhere in the middle when the nutrient concentration of mother and fetus became equalized (at around 50). A lesser percentage of the total nutrients would have been transferred (50 out of 100).

The human fetus undergoes a different process of implantation, however. The growing trophoblast dissolves the vascular walls of the blood sinuses that earlier developed in the endometrium. Individual sinuses coalesce into larger pools of maternal blood known as lacunae, or "lakes." Trophoblastic tissue extends down into these lakes like protoplasmic fingers bathing in a stream. Eventually blood vessels are developed in these trophoblastic fingers, and fetal blood will pulse through them as the fingers divide into profuse bushes of blood vessels. Only a single membrane of the fetal blood vessels will separate the maternal and fetal blood. This form of implantation, and the form of placenta that it gives rise to, called a haemochorial placenta, represent one extreme end of the range of placental forms that mammals produce, one that is shared by our closest primate relatives, the apes and Old World monkeys, and a few other more distantly related groups. It brings with it certain advantages, but certain disadvantages as well. We will consider these more closely in the next chapter. For now we can note that a principal advantage is a tremendous increase in the surface area over which maternal and fetal blood supplies come in contact. This results in a great increase in the total amount of nutrients that can diffuse from one side of the placenta to the other in a given amount of time, even when the concentration gradient is low. A principal disadvantage is that the maternal circulation no longer flows through narrow blood vessels so that the efficiency of the counter-current system is lost and new means must be employed to maintain blood flow and concentration gradients.

At the early stages of embryonic development these are not limiting problems. The successfully implanted embryo, bathing its trophoblastic tissue directly in maternal blood, has solved its second problem. Nutrients and oxygen begin to flow into its cells in seeming superabundance, and it begins to elaborate its tissues and to sculpt itself, following the age-old pattern of mammalian embryogenesis. But even before these tasks are begun in earnest, the embryo must solve its third problem. It must send a signal to its mother's ovary that will prevent the otherwise inevitable decline and destruction of the corpus luteum. Without such intervention, the corpus luteum will start to regress a week or ten days after ovulation and the levels of circulating progesterone will fall. It is these levels of progesterone that maintain the uterine lining. Without them the blood supply to the engorged endometrium will be pinched off and the outer layers sloughed in menstruation. The necessity for a constant supply of progesterone to maintain the integrity of the uterine lining has recently been underscored by the development of the so-called abortion pill RU486. This molecule acts to block progesterone receptors in a pregnant woman so that progesterone can no longer exert its effect. The result is abortion.

ET PHONE HOME

As soon as it has established a connection with the mother's blood supply, then, the trophoblast of the embryo begins to secrete a chemical signal, the molecule known as human chorionic gonadotropin. The name specifies that it is a molecule unique to humans, distinct even from the molecules that may have similar functions in apes and monkeys; that it derives from the placenta, and in particular from the part of the placenta that is produced by the embryo (the chorion); and that its function is to stimulate gonadal activity, just like the gonadotropins secreted by the mother's own pituitary. Indeed, this molecule, called hCG for short, is remarkably similar in its structure to LH, the pituitary gonadotropin that triggered ovulation and helped to create the corpus luteum in the first place. Both molecules are composed of two pairs of long protein chains, two alpha chains and two beta chains. The alpha chains in LH and hCG are identical in every one of their hundreds of amino acids. Only the beta chains differ, and only over a part of their length. The difference does not appear to affect the ability of the hCG molecule to bind to LH receptors. If anything it binds to them with even greater affinity than the LH molecule itself. The difference does, however, make the hCG molecule more resistant to chemical degradation in the body. While half of the LH circulating in the body is removed every thirty minutes or so, it takes nearly forty hours to remove half the circulating hCG. This molecule is therefore able to do what LH cannot: it can keep the corpus luteum alive and functioning, even helping it to increase its output of progesterone. The progesterone in turn staves off menstruation and maintains the integrity of the endometrium. By secreting this signal, the embryo saves itself.

This rescue must occur within a narrow time window as well, however. Once the process of detaching the endometrium has begun, it cannot be reversed. The embryo cannot afford to be shy. It cannot afford to whisper its message, it must shout. The message itself, hCG, is too large a protein molecule to pass through cell membranes. Even if the embryo began to secrete this molecule while traveling down the fallopian tube or traversing the uterus it would not reach its destination. The hCG has to be secreted directly into the mother's blood, and that can happen only after the embryo has successfully implanted and the trophoblast has broken through into the vascular lacunae of the endometrium. Only species that achieve such a direct connection between the embryonic and maternal circulation secrete such a signal: apes and Old World monkeys of course, but also a few less closely related species, such as horses, that achieve such a connection in certain limited regions of the placenta. Other species have other mechanisms to govern the life and activity of the corpus luteum. In many groups, such as rodents, it seems as if pregnancy is the assumed default for every reproductive cycle, with special mechanisms employed to destroy the corpus luteum in the event of failure. In humans it seems as if failure is the default assumption, with menstruation as the ordinary program. Special mechanisms must be invoked to shift this program for conception to take place.

It is the presence of hCG in the mother's bloodstream that is used today for the initial diagnosis of pregnancy, whether by an over-the-counter home pregnancy kit or by a clinical lab. In many cases hCG is detectable within days of implantation, about the time that a woman expects her next menstrual period or first suspects that it is late. But every kit and every clinical lab will insist that the test be repeated sometime later, two weeks or more after the woman's period was due, to be sure of the diagnosis, not because the test itself is so inaccurate, but because the fate of the embryo is far from certain in the early days of attachment to the mother.

THE CHOSEN FEW

A prospective study of 221 women in North Carolina who were attempting to conceive has provided the best estimate of the risk that a newly implanted embryo faces. The study participants collected urine samples at home that were later analyzed for hCG and compared with the clinical history of pregnancy diagnosis and outcome. Nearly a third (62 out of 198) of the pregnancies diagnosed by three consecutive days of detectable hCG failed before coming to term, two thirds of those (40 out of 62) were lost before the pregnancy was clinically recognized. On average menstrual cycles that ended in the loss of an undetected pregnancy were three days longer than nonconception cycles, but otherwise there was no outward manifestation that an embryo had ever been present. Of the 40 women who experienced this early pregnancy loss, 38 of them subsequently became pregnant again, and the success rates of their succeeding pregnancies were no different from those of the rest of the sample. Thus there is no reason to think that the women experiencing this early pregnancy loss were a special subset of the population. Rather it is reasonable to conclude that 20 percent of the embryos that initially begin implantation in most women are lost within a week to ten days.

The early days after implantation thus represent a concentrated period of risk for an embryo. Of those that survive long enough to be clinically recognized, another fraction will be lost in miscarriage and spontaneous abortion over the ensuing nine months. In the North Carolina study 14 percent of the clinically recognized pregnancies subsequently failed, a figure that is quite similar to the statistics on miscarriage and fetal loss found by larger regional studies in Hawaii, New York, and California. These studies also show that the risk of pregnancy loss continues to decline as the pregnancy advances, with the greatest risk occurring in the first month after the pregnancy is diagnosed.

The risks of failure prior to implantation are likely to be quite high as well, perhaps even higher than the postimplantation risks. There is very little reliable data to use, however, in estimating those risks. In 1959 a now famous study reporting on women undergoing elective hysterectomies was published by the Boston physician Arthur Hertig and his colleagues. One hundred and seven women voluntarily engaged in unprotected sexual intercourse around the presumed time of ovulation before their operations. The hysterectomies were performed between two and seventeen days after ovulation. When the uteri and fallopian tubes removed in these operations were carefully examined, thirty-four embryos were retrieved that could be grouped by developmental stage. From the twenty-four cases in which the hysterectomy occurred within a week of the presumed day of ovulation, eight embryos were recovered. Thus fertilization had been successful in at least one third of the cases. The actual fertilization rate might have been higher if some embryos escaped detection or had already been lost before surgery. Of the thirty-six cases where the hysterectomy was performed after implantation would have occurred (eleven to seventeen days after ovulation), embryos were observed in twenty-one instances, or 58 percent of the total. Although natural pregnancy loss was not observed in this study, Hertig did classify four of the eight preimplanation embryos he observed as morphologically abnormal and presumably nonviable, as well as six of the twenty-one postimplantation embryos.

Although Hertig's findings are fascinating, it is difficult for us to extrapolate very much from his data. The numbers are small and the actual events we are trying to tally, reproductive losses, were not observed. Fifty percent of the embryos observed in the preimplantation stage may well not have been viable, but we don't know when in the course of their subsequent development they might have been lost. Perhaps they represent the sort of embryo that would reach the early stage of implantation and begin to secrete hCG before failing. Perhaps they wouldn't even get that far. We can't be sure. Nor can we feel very confident about the 50 percent figure. It is based on a very small sample, and one that probably was incomplete. The higher recovery rate of later-stage embryos suggests that many preimplantation embryos may simply not have been recovered.

Another tantalizing study was conducted by an Australian researcher using early pregnancy factor (EPF), one of the local immune suppressants discussed earlier, as the basis for diagnosing the presence of a preimplantation embryo. That researcher detected EPF shortly after ovulation in eighteen of twenty-eight cases (64 percent) where women had engaged in unprotected sexual intercourse around the time of ovulation. Of the eighteen "conceptions" diagnosed by this method, fourteen (78 percent) never produced clinically diagnosed pregnancies. The EPF levels in these fourteen cases subsequently fell and menstruation occurred at the expected time. Taken at face value, these data seem to suggest that the overall rate of early embryo loss may be considerably higher than was indicated in the North Carolina study. But once again the sample size is small, and the interpretation of the EPF assay is uncertain.

Some researchers have attempted to fill the data void on the earliest period of embryonic life by resorting to mathematical models. The usefulness of this exercise depends, of course, on the validity of the assumptions that necessarily have to be made. The probability of eventual loss rises at an increasing rate the earlier in the pregnancy one begins, back to the first appearance of hCG. It seems reasonable at one level to assume that the probability of loss continues to rise back beyond that point to the moment of conception. Attempts to perform such an extrapolation, however, have produced widely varying results. Estimates of the probability of eventual loss for a newly fertilized egg range from 30 percent to 90 percent or higher for women in their mid-twenties. If one accepts the higher estimates, however, it's necessary also to assume that the probability of fertilization is virtually 100 percent every month, a rate that seems unreasonably high given our understanding of the factors governing the quality of eggs and sperm and the patterns of intercourse. Even under the controlled conditions of in vitro fertilization, where technicians bring eggs and sperm together in a petri dish—sidestepping all the uncertain timing of intercourse, variations in sperm count and concentration, and the hazards of the journey through the female reproductive tract—fertilization rates are much lower than 100 percent.

Although it may seem reasonable for modeling purposes to assume that the rate of embryonic loss continues to rise as one proceeds back from implantation to conception, it may not be true. One could argue that implantation is a crucial filter through which only viable embryos pass, and that the majority of embryos won't reveal their functional defects until that point. An embryo that is developmentally out of sync with the endometrium, for example, may be doomed from the start, but it's actual "failure" will occur when implantation is unsuccessful.

Although it is difficult to measure or estimate embryonic losses before the appearance of detectable hCG, the losses that occur after that point are remarkable enough. In the United States, it seems safe to conclude, roughly a third of the embryos that manage to break through to the maternal circulation and to begin secreting hCG fail anyway, two thirds of them before menses are delayed and a pregnancy is recognized.

IT'S WHAT'S ON THE INSIDE THAT COUNTS

There are few studies of non-Western populations with which to make comparison. The best study to date of gestational loss in a non-Western population is Darryl Holman's study of women in rural Bangladesh. The women that were his subjects were drawn from an area that has been under continuous demographic surveillance for decades, and their conditions of life are well documented. Most live by subsistence agriculture supplemented by fishing. Rice farming, which produces the primary staple food, is very labor intensive for both men and women. Few of the women have any formal schooling and most are married as teenagers. Their nutritional status tends to be quite low by international standards, revealed both by their depleted fat stores and by their low muscle mass. The incidence of diseases can be quite high, including highly debilitating endemic diseases like cholera.

Holman's study was also based on detecting hCG in urine and included information on 329 pregnancies. In contrast to the North Carolina study, however, the pregnancies were at various gestational stages when diagnosed and were followed for variable lengths of time. Combining all the information yields an estimate of embryonic loss of 34 percent for a gestational age comparable to that used in the North Carolina study. This figure seems remarkably close to the 32 percent found among the U.S. women, particularly considering the dramatic differences in ecology and lifestyle that distinguish these two populations.

It will be necessary to have a much larger set of studies from a broader range of populations and ecological circumstances before any conclusions can be carved in stone. Nevertheless, if we take the data available at face value, two questions arise: Why is the rate of embryonic loss so high? And why does it vary so little with environmental conditions? These two questions may have essentially the same answer: The embryos that are lost at such an early stage may be for the most part genetically nonviable. Hertig's study of recovered early embryos showed a high proportion of developmental abnormalities, and studies of recovered tissue from early miscarriages show much higher rates of chromosomal aberrations and developmental abnormalities than are observed at term. Many of these genetic defects result from the process of meiosis, during which the pairs of chromosomes are separated, with only one of each pair preserved in the mature reproductive sexual cell, or gamete, that is finally produced. Errors in this process can produce egg or sperm cells with an incorrect number of chromosomes, resulting in an embryo lacking a chromosome or having an extra one, or even an embryo with an entire extra set of chromosomes. These represent the gross genetic defects. Subtler defects are also a possible result of the process of sexual reproduction that combines maternal and paternal genes. Every gamete in a sense represents a different quasi-random pick of parental genes ("quasi-random" because genes that are close to each other on a given chromosome are "linked" and tend to stay together, while those far apart or on different chromosomes have a higher probability of being separated), and each fertilized egg is a random combination of those quasi-random selections. Not all of the resulting embryos receive equally viable combinations, and some inherit fatally defective ones. Evolutionary biologists have long been aware of this downside of sexual reproduction and have sought for evidence of compensating advantages, such as the ability to prevent pathogens from developing a deadly lock on predictable genetic combinations, or genotypes.

If nonviable embryos are unavoidable consequences of sexual reproduction, then there should be a tremendous selective pressure for exposing them and aborting the pregnancy as early as possible. Females of any species who spend long periods of time, not to mention large amounts of energy, gestating offspring doomed to perish are at a tremendous evolutionary disadvantage compared with females who reduce the amount of such nonproductive investment. The earlier defective embryos can be aborted, the less waste of time and resources. Thus it is predictable that most embryonic loss occurs early in pregnancy. Indeed, the delicate dance of implantation may even serve as a mechanism by which the genetic competence of the embryo is tested by the mother. The embryo must do its share of the work for implantation to succeed. If its abilities to respond to maternal signals and to generate signals in its turn are impaired, implantation is likely to fail. Perhaps as we learn more about the developmental genetics of this crucial period we will find that the process of implantation provides an important diagnostic assessment of the viability of the embryo.

Natural selection may produce mechanisms for the early exposure and elimination of nonviable embryos, but it may not be able to push the limit of such exposure much before implantation, since successful implantation represents the real initiation of maternal investment in the pregnancy, in terms of both time and nutrients. The adjustments in maternal physiology that occur after ovulation, in the luteal phase of the menstrual cycle, are not contingent upon the presence of an embryo, so the elimination of a defective embryo between fertilization and implantation would not achieve any metabolic savings. Nor would the elimination of embryos in this period hasten the next conception, so there is no time advantage to be gained. These considerations provide an additional theoretical reason for suspecting that the overall probability of embryonic loss may not be very much greater than the 33 percent observed at the time of implantation.

If early embryonic losses are predominantly due to the elimination of nonviable genetic combinations, then it would make sense that the rate of such losses would not vary much with environmental conditions. The rate of production of defective embryos should be predominantly a consequence of the roulette wheel of sexual reproduction, augmented perhaps by certain environmental agents that can cause genetic mutations. Without such external sources of pathology there is no reason to think that Bangladeshi couples, for example, would produce defective genetic combinations with a frequency any different from that of American couples. The close similarity between the rates of early embryonic losses in North Carolina and Bangladesh are, in fact, what one would predict if genetic defects were primarily to blame.

But isn't this a terribly inefficient way to deal with the problem of defective embryos? Wouldn't natural selection favor alternative solutions, such as producing more than one egg per month so that more than one embryo could attempt implantation? This is, after all, the way assisted reproduction clinics deal with the problem when performing in vitro fertilization (IVF). A woman is induced to produce many eggs, dozens even. They are exposed to sperm, and some of them are successfully fertilized. Of that subset a few, often three but sometimes more, are chosen on the basis of their progress through early cleavage stages and introduced to the uterus to attempt implantation. By returning more than one embryo to the uterus, the clinicians raise the probability that at least one will be successful. If this improves clinical success rates, why wouldn't an analogous physiological process raise natural success rates and hence be favored by natural selection? The answer is perhaps also apparent from the experience of IVF clinics. Introducing more than one embryo into the uterus raises the probability that at least one will successfully implant, but it also raises the probability that more than one will implant. The multiple pregnancy rate in IVF currently presents a moral dilemma to the medical community. Physicians must balance their patients' often desperate desire for pregnancy and the clinic's need to have a success rate that is competitive with other clinics against the known medical risks of multiple pregnancies and the attendant social and psychological challenges for the parents.

In terms of natural selection the problem is not a moral one, however, but simply an actuarial one. Although it is difficult to be sure, it is likely that until relatively recently the cost of a multiple pregnancy was too high a price to pay. Rates of prematurity and low birthweight increase markedly for multiple pregnancies even in industrialized societies. Under more demanding ecological circumstances and in the absence of modern obstetric and postnatal medical care, these rates and the attendant risk of mortality to infants and mothers would presumably have been higher still. Selective abortion or resorption of all but one implanted embryo is also much more difficult in humans than in other species, where multiple pregnancies are the rule and implantation is much more superficial.

Human reproductive physiology appears designed, then, to run the narrow course between these competing risks: the risk of wasting effort on a defective embryo and the risk of expending excessive effort on a multiple pregnancy. The solution has been to attempt only one pregnancy at a time (usually) and to develop mechanisms to filter out defective embryos as early as possible relative to the schedule of investment. Of course, multiple pregnancies do occur naturally in humans, but at low rates. Those rates show some predictable patterns as well. The rate of monozygotic twinning (one fertilized egg that cleaves into two embryos at an early stage) appears to be highly heritable and relatively invariant. Dizygotic twinning, however, resulting from the production of more than one egg, increases predictably with maternal age. This is ordinarily understood as a consequence of the falling levels of estrogen hormones in older women and a compensatory response that leads to more than one follicle developing to the point of readiness for ovulation. We will revisit these age-related changes in ovarian function in a later chapter. But it is consonant with evolutionary logic that natural selection might weigh the risks of multiple pregnancy against the savings in time waiting for a successful embryo to implant differently in older and younger women.

MAKING THE COMMITMENT

The high rates of early embryonic loss and the consistency of these rates across different populations and ecological circumstances can both be understood from an evolutionary perspective. After implantation, however, embryonic and fetal loss rates still show less ecological variability than we might expect. There are certainly environmental circumstances that raise rates of spontaneous abortion, particularly acute diseases such as malaria or behaviors such as smoking that can adversely affect placental function. But other conditions, such as malnutrition and famine, do not seem to raise abortion rates significantly. Even under the most severe circumstances, such as the "Dutch hunger winter" during the Nazi occupation, gestation lengths shorten slightly and rates of low birthweight increase, but not rates of spontaneous abortion. This presents a challenge to the evolutionary perspective. We will see abundant evidence later for reductions in the probability of conception under adverse energetic conditions, and the argument will be made that this represents functional avoidance of investment in reproduction when the probability of success is diminished. Wouldn't the same logic lead one to expect that an energetically stressed mother should be more likely to abort her fetus and thus avoid continued investment in an offspring with diminished prospects of survival?

Perhaps. But two arguments can be made that suggest that the threshold for facultative abortion should be much higher after implantation than the threshold for avoidance of conception. David Haig has pointed out that the genetic interests of an offspring are, like its genes, different from those of its mother. Things that are good for the spread of the offspring's genes are not necessarily good for the spread of the mother's genes. On this basis he predicts that offspring will be selected to use their own physiological means to try to extract more investment from the mother than is optimal for her. Implantation allows the embryo to secrete hormones into the mother's circulation that bias the distribution of metabolic energy toward the fetus even to the point of compromising the mother's health. Such a metabolic "conflict" between the mother and the fetus may also make it more difficult for the mother to abort the pregnancy when the cost/benefit ratio becomes unfavorable from her own perspective.

On the other hand, such an appeal to metabolic conflict between mother and offspring may not be necessary. The fact of successful implantation alone may tip the scales of the cost/benefit equation significantly since it implies that the embryo has passed a significant "viability" test. The probability that a given pregnancy will be successful increases by at least 33 percent after implantation, and the "cost" necessary to tip the scales away from further investment becomes equivalently higher.

These two arguments are not mutually exclusive, of course. They may, in fact, reinforce each other. In any case, implantation appears to represent the crossing of the Rubicon for further maternal investment. Once an embryo has made it that far, maternal physiology seems to favor its continued survival even under adverse conditions.

A CURSE OR A BLESSING?

All human eggs do not end up as pregnancies. Of those that are ovulated, only a fraction will be fertilized. Of those that are fertilized, only a fraction will successfully implant. Because not all ovarian cycles end in pregnancy, many end in menstruation, the shedding of the outer layer of the endometrium and the attendant vaginal discharge of blood and tissue. Recently this phenomenon itself has been viewed as an evolutionary puzzle deserving of a solution. In 1993 Margie Profet pointed out that menstruation in women had been largely taken for granted as a physiological necessity rather than appreciated as a potentially adaptive response to particular ecological circumstances. Not all female mammals menstruate, she pointed out, and there are a variety of mechanisms for dealing with the uterine lining when pregnancy does not occur. In some the lining is resorbed; in some it is preserved for a subsequent potential pregnancy. Only in certain species, particularly Old World primates, is the outer layer of the endometrium actively detached and shed externally. Why should a mechanism so wasteful of tissue and blood evolve?

Profet suggested an answer that she said came to her in a dream: Menstruation occurs to cleanse the female reproductive tract of potentially harmful bacteria that may have penetrated the physiological defenses discussed above by "piggybacking" a ride on sperm cells. The nutrient-rich, immunologically suppressed environment of the endometrium would serve as a wonderful "petri dish" for the propagation of such bacteria, which might then, by way of the fallopian tubes, gain access to the woman's body cavity, leading to systemic infection. Menstruation would provide a mechanism for disposing of such a bacterial culture. Once Profet had formulated this hypothesis, she viewed many of the features of the endometrium and the process of menstruation as designed to achieve this specific purpose. For example, the peculiar spiral arteries that develop in the endometrium (of which we will hear more in the next chapter), which cause blood to enter the endometrial lacunae in pulsing jets, could be specially designed to jettison the endometrium in menstruation and to "pressure-wash" residual bacteria away. If menstruation evolved to serve this function, Profet argued, its distribution across species should be affected by the risk of infection. The risk of infection in turn, she argued, should be affected by the rate of copulation and perhaps the number of different partners. Using this argument, she tried to map the distribution of menstruation against the mating systems of different mammals. Her attempt was very problematic, however, since it led her to redefine menstruation to include any fluid discharge from the female reproductive tract, such as the thin vaginal spotting that many dogs display when in heat. Physiologically, though, this phenomenon is very different from menstruation in women. A physiologist would refer to it as diapedetic bleeding, fluid leakage as a consequence of the engorgement of the reproductive tract at midcycle, which is very different from the physical detachment of the outer endometrium at the end of the cycle.

Whatever the weaknesses in her argument, Profet did focus the attention of other researchers on the phenomenon of menstruation and its possible functional significance. Why indeed such a wasteful, and for some uncomfortable, discharge of tissue and blood? The blood loss alone puts women of reproductive age at a significantly higher risk of anemia than other age-sex classes. Why don't humans resorb the endometrium or reuse it like other species?

A number of alternative hypotheses were advanced in Profet's wake, along with many criticisms of her formulation. One proposal noted that menstruation also serves to remove embryos that may have initiated implantation only to fail. It is not clear, however, why failed embryos need to be removed, or why resorption wouldn't be equally effective. Another argument notes that the human endometrium, like that of other Old World primates but unlike that of many other mammalian species, differentiates irreversibly into secretory tissue before implantation occurs, making it impossible to reinitiate a new developmental sequence with the same tissue. It is as if the ship has to be launched before the passengers are on board. If the passengers do not "jump aboard" at the crucial moment, the opportunity is lost. The ship cannot be recalled to the dock. Instead, that ship must be scuttled and a new ship prepared for launching. This scenario captures important features of human endometrial development, but itself begs for explanation. Why must the endometrial "ship" be launched so early in some species and not in others? And even if it can't be reused, wouldn't it be more efficient to resorb the material of the failed endometrium?

The most powerful attempt to address all these issues was made by Beverly Strassmann, who also provided the most thorough critique of Profet's hypothesis. Strassmann suggests that the reason menstruation occurs is simply that other alternatives are too costly in metabolic terms. The outer layer of the human endometrium, she argues, is too bulky to be efficiently resorbed, as it is in species with a much thinner outer layer. And to maintain the endometrium in a state of readiness for another month, awaiting the next opportunity for implantation, is calorically more expensive than simply constructing a new endometrium from scratch. To support her argument she notes the elevation of a woman's metabolic rate during the second half of her menstrual cycle, when her endometrium is in the secretory phase of its development, compared to the rate during the first half of her cycle, when the endometrium is in its proliferative phase. Since spending two weeks in the proliferative phase is manifestly cheaper, she argues, than spending an additional two weeks in the secretory phase, natural selection will favor endometrial sloughing and regrowth over endometrial maintenance. She argues that endometrial cycling is analogous in this way to other types of reproductive quiescence, such as the annual regression of reproductive organs in seasonally breeding species, that are easily understood as metabolically efficient. She distinguishes her argument from the terminal differentiation argument against reuse of the endometrium described above. That argument appears to her to depend on an unexplained constraint—that the endometrium is incapable of reuse—in a way that renders it weak in evolutionary terms.

However, Strassmann's own argument depends on constraints, even though they may not be explicitly acknowledged. For example, why can't the secretory activity of the endometrium be suspended while it awaits the next opportunity for implantation? That might achieve a sufficient savings of energy to make maintenance of a quiescent endometrium an efficient alternative. But there is an even more interesting constraint assumed by Strassmann's argument: that the next opportunity for implantation won't occur for two weeks. What if a new implantation opportunity were only a week away? Or a few days? Would it then become more energetically efficient to wait?

Sometimes obvious questions are overlooked, and sometimes they prove illuminating when they are finally asked. Profet pointed out that the question "Why do women menstruate?" had not been asked in a thoughtful way. But asking that question has now led us to another: Why does ovulation occur only once a month? This does, after all, place a limit on human fecundity, with maximally twelve or thirteen opportunities for conception per year for a given woman. A mouse will have six or seven opportunities for conception for every one that a woman has.

Why do women ovulate so infrequently? Some alternatives are conceivable, if hypothetical. Follicles and the eggs they contain could mature at intervals more closely spaced than a month, so that newly ovulated eggs appeared every few days. Male sperm maturation is an extreme version of such overlapping gamete maturation, with new mature sperm produced in prodigious quantities nearly continuously. Without going to that extreme, females of other species manage to have overlapping waves of follicular maturation, so that a new crop of oocytes is maturing toward ovulation even while the previous cohort is on its way down the fallopian tubes toward the uterus and possible fertilization.

But let us assume for the moment that there is some technical difficulty with engineering overlapping gamete maturation in human females. We can still ask why it takes two weeks to create another conception opportunity after a prior opportunity has been lost. Two possible constraints suggest themselves: first, it takes that long to mature another egg; second, it takes that long to mature another endometrium. Strassmann's argument, though she doesn't present it in these terms, essentially embraces the first of these constraints, while the terminal differentiation argument essentially embraces the second. Given that you're going to have to wait two weeks for another egg, Strassmann argues, there's no point in maintaining the endometrium. A rephrase of the terminal differentiation argument might state the case differently: given that you're going to have to wait two weeks for another endometrium, there's no point in producing another egg.

It's difficult to know which of these two constraints is the operant one in humans. But a sideways glance at other primates may provide a clue. The twenty-nine-day ovarian cycle that is typical of humans is not by any means a constant feature of all our primate relatives. Apes and Old World monkeys tend to have ovarian cycles that cluster in the twenty-five- to thirty-five-day range, but New World monkeys, in contrast, often have cycles that are much shorter, in the range of twelve to twenty days, characterized by shorter follicular phases than in the Old World monkeys. These monkeys, however, have a more superficial form of implantation and placenta development that is associated with a thinner endometrium. They also do not menstruate, as is typical of the Old World monkeys. It would seem from their example that it is quite possible to mature a viable primate egg on a more rapid timetable than two weeks. Instead it would seem to be the time required to produce the peculiarly thick endometrium of the Old World primates, associated with their peculiar form of implantation and placentation, that constrains the periodicity of the cycle.

Strassmann's argument is thus not at all free from dependence on constraints. Menstruation in women occurs, she would have it, because it is cheaper than maintaining the endometrium for another two weeks, *and because another egg can't be produced sooner.* As an alternative, we have the terminal differentiation argument. Menstruation occurs in women because the form of endometrium necessary to support the peculiar implantation and placentation of humans requires terminal tissue differentiation before implantation, precluding the reuse of the endometrium after the window of opportunity for implantation has passed. Growth of a new endometrium is necessary, *and that takes two weeks*. In agreement with Strassmann, the terminal differentiation hypothesis assumes that the old endometrium is too bulky to be resorbed efficiently.

Both of these hypotheses are still alive and well. But the terminal differentiation hypothesis points in an interesting direction toward the special characteristics of implantation and placentation in Old World primates that humans share. These characteristics will occupy more of our attention in the next chapter. Yet we should remember that menstruation would never be necessary at all if conception never failed. It has been pointed out by many people that menstruation is a much more common experience for women in developed Western societies than for women in traditional societies or, presumably, for women in prehistory. Among the latter two groups a greater percentage of a woman's reproductive life might be spent in nonmenstruating conditions of pregnancy and intensive lactation. But even in the most remote and traditional of contemporary societies menstruation is a common enough experience for cultural elaboration and ritual to develop around it. Strassmann's own work among the Dogon of Mali, for example, has centered on the ritual separation of menstruating women in special menstrual huts.

Conception failure is a common feature of human reproductive physiology. At first this feature seems surprising, if not alarming. An evolutionary perspective, however, enables us to understand that such a high failure rate may be a necessary consequence of sexual reproduction. Given this likelihood, human physiology seems elegantly designed to minimize investment in nonviable embryos and to generate a new conception opportunity in the shortest possible time. One significant constraint in generating that new opportunity may be the production of a new endometrium. Once an embryo has made it through this minefield, however, the maternal commitment to the pregnancy seems to increase. The high tide of mortality risk subsides, for a while.

A TIME TO BE BORN

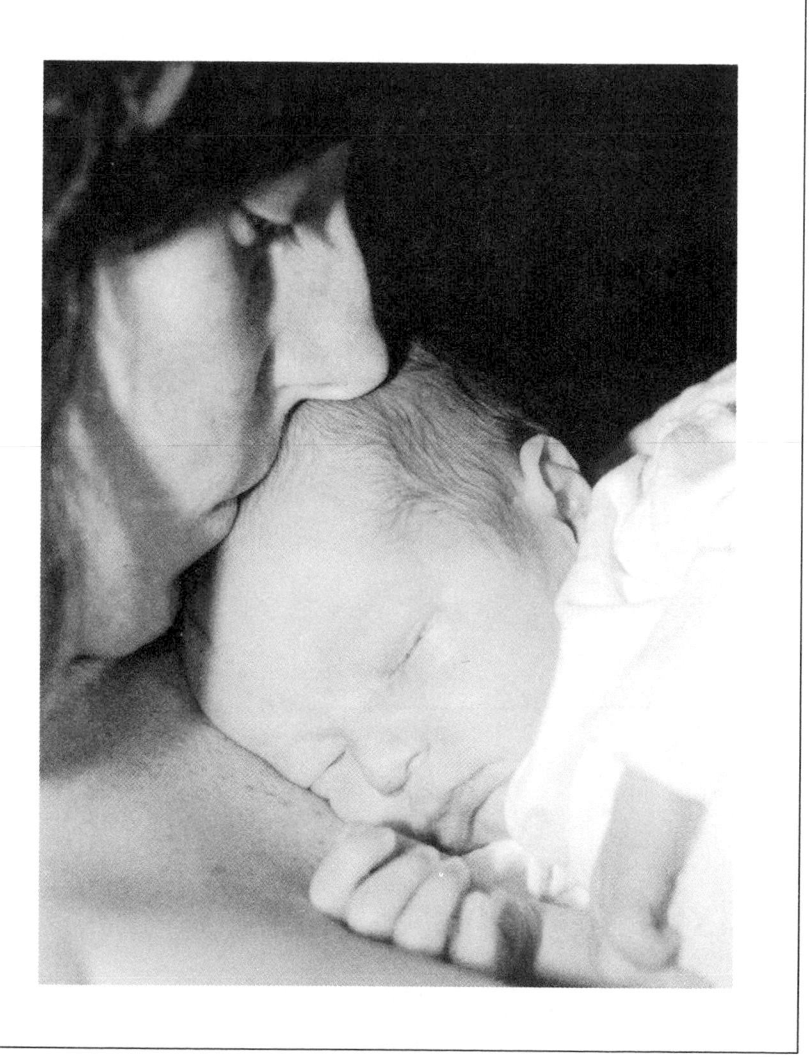

AFTER THE DRAMA OF IMPLANTATION, birth is the most dangerous threshold a new human being has to cross. The ninety-six hours or so following the onset of parturition constitute the greatest single period of mortality risk that the typical human will ever face. Moreover, there is reason to think that the course of human evolution has intensified this risk rather than abated it. The most obvious evidence of this is the unusual difficulty women have in labor and delivery compared with other mammals and even other primates. Most pregnant female primates seek solitude when giving birth. Primate labor, when it has been observed, does not seem excessively difficult. The female may pace and change positions frequently, but then usually delivers in a squatting position without assistance, often delivering the placenta herself by pulling on the cord or allowing it to fall out later as she moves about. She often begins to walk around soon after delivery, sometimes holding her new baby with one hand if it is not yet able to cling to her body on its own.

The contrast with human delivery is quite striking. While primate birth is usually a private experience, human birth is usually a social one. In every human culture birth normally occurs in the presence of others, usually women who have experienced childbirth themselves. Labor can be prolonged, manifestly painful, and often exhausting. Even after natural, unmedicated deliveries, mothers often sleep for much of the ensuing twelve hours. And these are the easy cases. There are cases of difficult presentation that may require the intervention of others for a vaginal birth to occur. At the extreme are cases of cephalopelvic disproportion (a fetal head too large to pass through the mother's pelvis) and other causes of arrested delivery, situations that can lead to the death of the mother as well as the infant if there is no radical intervention.

WITH A LITTLE HELP FROM HER FRIENDS

The social context for human birth is so important that it appears mechanisms have evolved to forestall delivery if social support is not available. The first scientific demonstration of this fact came about almost by accident. A team of researchers led by Roberto Sosa were interested in the impact of social support during labor and delivery on the quality of maternal-infant social interactions after delivery. Mothers who went through labor in a more sociable context, they reasoned, might be more apt to interact socially with their own newborns than those who were deprived of social interactions during labor. To test this hypothesis they made use of an institution in Guatemala known as the doula. A doula is someone who attends a mother during labor, not a midwife, not a birth coach, not anyone with special knowledge or expertise, not even necessarily someone known to the woman delivering. Just a companion who remains with the mother throughout the time of her labor, keeps her company, talks to her, gives her the reassurance of another human presence. Sosa and his colleagues set up a study of first-time mothers coming to a Guatemala City hospital to give birth. The participants were randomly assigned either to have a doula or not. Those without doulas experienced the same conditions for labor and delivery, but rather than enjoying constant human companionship they spent the majority of their time before the birth alone, with nurses and doctors making occasional visits to check on the progress of their labor.

The plan for the study was to have twenty uncomplicated, vaginal births in each of the two groups (doula and non-doula) and to compare various measures of maternal-infant interaction after birth. In order to reach the target sample size, however, the researchers had to recruit extra women to allow for a certain frequency of complicated deliveries. To reach their target they eventually recruited 33 women into the doula group, but had to recruit 103 women into the non-doula group to end up with twenty uncomplicated births. Contributing to this difference was a Caesarean rate nearly half again as high in the non-doula group as in the doula group (27 percent versus 19 percent) and an average length of labor more than twice as long (19.3 hours versus 8.8 hours). Not surprisingly, perhaps, the social interactions between mothers from the non-doula group and their infants were considerably more listless and infrequent than those of the doula group with their infants. The non-doula mothers also spent a greater portion of the postpartum study period asleep.

It is quite likely that the random assignment of women to labor in solitude contributed to the slow progress of labor and the high rate of complications. Psychological anxiety leads to the release of chemicals into the mother's blood stream such as epinephrine (also known by its British name of adrenaline) that have a powerful effect in arresting uterine contractions. The presence of a sympathetic human being may go a long way toward reducing the panic that a woman may feel, especially approaching the experience of childbirth for the first time, releasing this emergency brake on the progress of labor. Given the potential importance that the intervention of others may have in the success of human birth, the existence of this brake and its sensitivity to the presence of others may be very adaptive. The anxiety that a woman who begins labor without social support feels may help to delay its progress until such support is available. Mind and body may work together in this way to ensure that parturition occurs under the most favorable conditions.

It is particularly ironic and appalling to realize that medical practice in our own society for a long time required women to go through labor under "non-doula" conditions, bereft of friends and family in unfamiliar hospital surroundings. Before delivery became the province of male practitioners in England in the eighteenth century, a woman spent not only her labor but much of the preceding period of her "confinement" in the company of women friends, her "gossips" or "God sibs," who provided constant human companionship (as distinct from the midwife, who might manage the delivery itself). This practice is very reminiscent of the customs of many traditional societies. Yet only in recent years have American hospitals allowed family members or friends into the labor and delivery room, thus restoring a crucial aspect of the experience of childbirth to which our physiology has probably been adapted.

CRUNCH TIME

But why do we need this social support during parturition that our primate relatives seem to do without? Why is it so important for other people to be in attendance at a human birth that a woman's body will attempt to slow down the process in their absence? Why, in short, is human birth so difficult and dangerous? One answer seems obvious to anyone who has contemplated an illustration of a full-term fetus inside a woman's body. The baby is too large to manage an easy exit. Indeed, the more one contemplates the size of the baby's skull relative to the size of the mother's pelvis, the more miraculous any natural delivery appears. In relative scale the feat of passing a full-term fetal head through a woman's pelvis is comparable to swallowing a baseball. The passage of a typical monkey baby through its mother's birth canal is nowhere near so difficult because the fit is nowhere near so tight as it is for a human. Nor is the monkey baby obliged to twist its head relative to its torso in passing through its mother's pelvis as a human infant is.

Although a baby may have several skeletal dimensions that seem to pose a challenge to an easy delivery, the head is the main problem. On the mother's side, the internal diameters of the lower, or "true," pelvis are the major constraint. So tight is the fit on average that special hormones are released late in gestation to help loosen the ligaments binding the halves of the pelvis together. During delivery these softened ligaments can stretch to increase the diameter of the birth canal slightly by allowing the pelvis to become partially disarticulated. Not every birth poses the same difficulty, however. In some cases of particularly wide pelvises and particularly small infant craniums the head may slip out with scarcely a pause. In other cases passage of the infant head through the maternal pelvis is physically impossible. Nowadays a Caesarean delivery can be performed in such cases with minimal risk. But before the advent of modern anesthesia, antibiotics, and sterile procedures, Caesarean deliveries were extremely perilous. Before male practitioners took over routine deliveries in England, doctors were primarily called in only to deal with cases where a natural delivery was deemed impossible. They brought a gruesome toolkit of implements with which to puncture and crush the infant's skull and hooks to use in drawing out its corpse, all in the hopes of saving the mother's life.

How did our species end up in this situation? There are two obvious escape routes for evolution to follow: toward a smaller full-term fetal head or toward a larger maternal pelvis. Why hasn't evolution moved us in one direction or the other? A smaller fetal head might, of course, sacrifice other important aspects of our biology, assuming that our three- to fourfold increase in cranial capacity over our closest primate relatives has some relationship to our cognitive and linguistic abilities. This increase in brain size is primarily a function of increases in the neocortex, particularly in the prefrontal lobes, set in train early in fetal development. Brain growth then proceeds rapidly through fetal life and infancy and is virtually complete in early childhood. Early brain development is a standard mammalian pattern and probably necessary in order to establish the capacities of the central nervous system as quickly as possible. If it is not possible to postpone brain growth to a later period in life and if it is advantageous to retain our adult brain size, then the only other way to achieve a smaller full-term fetal head size would be to shorten the gestational period. In some sense we may already have pushed things pretty far in that direction. Although human gestation is only a month and a half longer in absolute terms than chimpanzee gestation, human birth occurs significantly earlier relative to the trajectory of brain growth. Compared with other primate infants, human newborns are particularly feeble and helpless, unable even to hold up their heads or to find the nipple without assistance. Premature human infants face a number of increased risks stemming from the immature state of their central nervous systems even with modern medical care. It seems unlikely that evolution could have advanced natural parturition any earlier relative to brain growth among our prehistoric ancestors without greatly raising the probability of infant death.

What about the other escape route? Why hasn't evolution provided us with a larger pelvis? The fact is it has, at least for half of us. Of all the bones in the human skeleton the pelvis shows the greatest degree of sexual dimorphism, or difference between males and females. Even first-time students can be taught in a matter of minutes to determine the sex of a human skeleton from the pelvis with reasonable accuracy. Other bones may show slight differences between the sexes on average, but the normal ranges of variation usually overlap to a considerable degree. In contrast, the female pelvis shows evidence of natural selection for an increased birth canal diameter, a feature that is absent in the male pelvis. Without this dimorphism, if a female had a pelvis of the male form and size for her height, no full-term human fetus would ever be born vaginally. In particular, the pubic bones at the front of the pelvis are proportionately longer in females, giving the top of the true pelvis, or pelvic inlet, a more circular and less heart-shaped outline, while the ischial bones and the sacrum at the bottom of the true pelvis, or pelvic outlet, have spread farther apart, increasing the diameter of the outlet and making it more circular in form as well. The large iliac blades that rise above the true pelvis also tend to flare out more in females than in males, but this has no direct relationship to the process of parturition.

The changes in shape and relative dimensions of the true pelvis in females are just sufficient to allow a full-term fetal head to pass, but only just. Even then the pelvic ligaments may need to stretch considerably, the fetal cranium may undergo significant compression of its as yet unfused bones, and the fetal head will have to undergo both extension (bending back at the neck) and torsion (rotation through nearly ninety degrees) in order to travel the eight inches through the pelvic birth canal. And even so there are still a significant number of cases in which these changes in pelvic size and form are not enough and the fetal head is unable to pass through the pelvis. Evolution seems to have been particularly stingy, endowing human females with a pelvis just large enough for a baby to be born with extraordinary difficulty and considerable pain most, but not all, of the time.

WALK THE WALK

But why so stingy? Why not one more inch in the radius of the birth canal, an inch that could make all the difference, making parturition less deadly and frightening and traumatic? Surely there would be a selective advantage to individuals who carried a genetic disposition toward a larger female pelvis: fewer prolonged or obstructed labors to endanger both mother and infant. The answer, many suspect, is that the obstetric benefits to be gained from a larger pelvis are counterbalanced by mechanical costs in relation to another key function of the pelvis: support for bipedal locomotion.

Our distant ancestors stood up and walked before they developed a human-sized brain. In the rock of East Africa are preserved the footprints of australopithicines, the prehistoric apes closest to the human lineage, who walked with a human-like, two-footed gait across the mud of a receding lake bed. The bones of these creatures display a pelvis that has changed dramatically from that of other apes, such as chimpanzees and gorillas: bowl-shaped rather than long and narrow. The iliac blades flare out as much to help support the abdominal organs in an upright posture as to provide attachment surface for muscles. But the muscle attachments have also shifted, as has the shape of the femur, or thighbone. An animal that travels on all fours, even a knuckle-walker like a chimpanzee or gorilla, does not have the same balance problems that a biped does. When the right hindfoot lifts, body weight is usually shared between the left hindfoot and the right forefoot or hand (unless the animal is in rapid forward motion, when balance is not the issue as much as propulsion). When a biped lifts one foot its weight must be carried on the other and the upper body must be shifted in order to maintain the center of gravity over the weight-bearing foot to prevent the creature from toppling over. To accomplish this efficiently, our feet are positioned more under the center of our bodies than our hips are. Our thighs angle inward and meet our lower legs at the knee in a slight angle. Although we may occasionally stand with our feet spread apart for increased stability, especially when lifting or pulling a load, we walk and run with our footsteps falling nearly in a line. Walking with feet spread is exceedingly awkward, forcing us to waddle along, shifting our center of gravity a long way side-to-side with each step, as anyone knows who has used old-fashioned, bear-paw snowshoes. A lot of energy is dissipated in this waddling motion. As a result it takes more energy to walk the same distance in snowshoes than in sneakers, even without the snow. Our locomotory apparatus from the pelvis down shows considerable evidence of efficient design for bipedal locomotion, efficiency wrought by natural selection. Much of that efficiency of design was already present in the australopithicines who left their footprints in the East African mud.

But then a new selection pressure was brought to bear on the pelvis of females in the evolving lineage of humans: increased fetal head size. This problem was exacerbated by the new orientation of the pelvis with its shorter, compressed form and consequently constricted birth canal, compared with the older, quadrupedal primate form. Natural selection began to modify the female pelvis toward the modern human form, increasing the critical diameters of the birth canal. But as it did so the hip joints where the femurs meet the pelvis moved farther from the midline of the body. This change effectively increases the length of the lever on which the weight of the body must be balanced when walking bipedally, so that greater force is needed to balance the same weight on the other side of the fulcrum point, in this case the hip joint. To help compensate, natural selection also acted to increase the length of the lever arm of the hip joint. This is accomplished by increasing the length of the neck of the femur separating the femoral head that fits into the socket of the pelvis from the site of attachment of the muscles responsible for tilting the pelvis slightly to pull the center of gravity toward the supporting limb. To maintain the mechanical advantage of these muscles the site for the attachment of their other ends, the crest of the iliac blades of the pelvis, also had to be extended outward (hence the greater flaring of the iliac blades in female than in male pelvises). Even with all these changes to compensate for the outward displacement of the hip joints caused by the increase in the diameter of the birth canal, the female's pelvis must be tipped farther than the male's at each step in order to maintain balance.

These changes all take their toll on the efficiency of bipedal locomotion and balance. Athletic activities that emphasize these functions of the lower skeleton usually end up "selecting" for females with narrower pelvises. The elite levels of track, gymnastics, or ballet are usually occupied only by women at the narrow extreme of the normal range in pelvic breadth. This is a different issue from muscular strength and coordination. Competitive female swimmers, for example, do not have particularly narrow hips, since the mechanical principles that govern their sport are very different from those that govern the strict efficiency of bipedal locomotion.

One might still wonder, however, whether a slight decrease in bipedal efficiency would not be a price worth paying in order to ease the difficulty and danger of childbirth. Would a slight decrease in bipedal efficiency really have lowered the survival probabilities of our female ancestors more than a larger birth canal would have raised them? It may seem unlikely, but we should not underestimate the costs of further reducing bipedal efficiency. Contemporary hunter-gatherers—people who still live as our Pleistocene ancestors did, as itinerant foragers of wild plants and animals—routinely travel impressive distances on foot, often carrying children, food, water, firewood, and other considerable burdens. Anything that increased the frequency of lower limb muscle strains or sprains, pulled or torn ligaments, let alone fractures or dislocations, could render an already demanding lifestyle even more difficult. Anything that increases the caloric cost of bipedal activities will intensify the challenge of consuming enough calories to balance expenditures. And while the risks of childbirth may be great, they are faced only a discrete number of times in a woman's life. Walking and carrying are unavoidable daily activities.

At some point, then, the benefits of a still larger pelvis diminish while the costs increase. A balance of these opposing selection pressures appears to be struck when the female pelvis is just big enough for birth to be successful most of the time, with appropriate social support, but no bigger. But in this coevolutionary process it is fetal head size that places constraints on maternal pelvic dimensions, not the other way around. What is it that determines the size of the full-term fetal head? I noted above that this is roughly equivalent to asking what determines the length of gestation and that human infants are particularly altricial, or underdeveloped, at birth. Independent of the selective pressures working on the mother's pelvis, wouldn't there be selective pressure to prolong gestation somewhat in order for the infant to be more robust and fully developed at birth?

PULLING THE PLUG

Babies can survive when born prematurely, though their survival chances diminish rapidly the more immature they are. All the organs of a fetus have been formed by the end of the second trimester; the skeleton has been laid down in cartilage and has begun to ossify; brain stem functions have been established and sensory responses can be demonstrated, indicating functional cortical brain circuits. The baby is basically all there two to three months before its due date. The final maturational events in some organ systems, notably the lungs, are postponed until just before birth. But nothing essentially "new" is created during the last third of gestation. The last trimester is instead devoted to preparing for birth: the fetus grows in scale and bulk, especially in the brain, and accumulates important energy reserves in the form of fat. Fat reserves in the fetus normally accumulate faster than other tissues during this period. The growth of the brain also requires increasing amounts of fat as individual nerve projections begin to be coated with myelin, the fatty sheath that surrounds and insulates them.

The more fat a fetus can accumulate during its final trimester the greater its chance of survival as a neonate and an infant. Birthweight is an extremely powerful predictor of infant mortality around the world, and babies that are born below 2,500 grams, a condition that implies scanty fat reserves, are at particularly high risk. But this fact presents something of a paradox. Why are low birthweight babies born? Why aren't they kept in the womb a little longer, until they can accumulate enough fat to have a fighting chance? Rather than resulting from a prolonged gestation, underweight babies tend to be somewhat premature, and both conditions are more likely when the mother is undernourished. Why under these conditions should a baby be born in an even less mature state than usual?

One explanation with intuitive appeal that might account for the limitation on gestation length, the possibility that placental function simply can't be extended much beyond nine months, turns out not to be true. The longest well-documented human gestation, reported in the respected British medical journal *The Lancet,* is over twelve months. The baby that was delivered turned out to be anencephalic, missing most of the neocortex of its brain owing to a developmental defect. Yet however pathological the fetus, the case (and many others like it) clearly demonstrates that there is no inherent limit on placental function at nine months. The placenta is perfectly capable of continuing gestation if nothing happens to initiate parturition.

What does control the initiation of parturition, then? This turns out to be something of a continuing mystery. At one point it was supposed that the mother's physiology was in charge of deciding the moment for labor to commence. Perhaps stretch receptors in the uterus register maximal distension and initiate a cascade of signals to start labor. That would help account, for example, for the tendency of twins and triplets to be born prematurely. A curious observation made on domestic sheep, however, called this scenario into question. It was observed that pregnant ewes that fed in a field containing a certain species of pigweed *(Veratrum californicum)* never initiated labor. They continued to gestate their fetuses way beyond normal term, often with fatal consequences for mother and offspring. Autopsies revealed that the fetuses had failed to develop a functional pituitary gland, a pathology eventually traced to the pigweed in the mother's diet. The suspicion was raised that this developmental defect was somehow responsible for the failure of parturition to occur normally, suggesting that the fetus, rather than the mother, might determine when it was time to be born.

Based upon this clue and the subsequent research that it stimulated, a quite satisfying account can now be given of the events triggering parturition in sheep. To understand this drama an introduction to a few of the leading physiological "players" is in order. As is the case with humans, pregnant sheep produce large amounts of progesterone. In addition to maintaining the integrity of the uterine lining, progesterone reduces the contractility of the muscular part of the uterus, the myometrium, thus impeding the development of any contractions. Other hormones have the opposite effect, exciting contractility of the smooth muscle fibers of the uterus. Among the group of contraction stimulators are the steroid hormone estradiol and members of a group of physiological regulators known as prostaglandins. Estradiol does not directly stimulate uterine contractions so much as it facilitates the action of the other hormones by counteracting the sedative effects of progesterone, by promoting the development of connections between the smooth muscle cells of the uterus (gap junctions, which are necessary for coordinated waves of contraction), and by stimulating the production of prostaglandins and their receptors. Prostaglandins are very powerful stimulators of smooth muscle fibers. As a class they are derived from certain fatty acids and are usually produced near the tissues they stimulate. They contribute to the control of vascular tension by stimulating contraction of the smooth muscle fibers surrounding blood vessels. Overproduction of prostaglandins can contribute to headaches by causing vasodilation or constriction of cerebral blood vessels. Among the key actions of aspirin is the inhibition of prostaglandins, hence its use as a headache remedy. Two other hormones that play important roles in the drama of parturition in sheep are the pituitary protein hormone adrenocorticotropic hormone, or ACTH, and the adrenal steroid hormone, cortisol. Cortisol is an important regulator of blood sugar in sheep and humans, helping to maintain an adequate blood level of this basic metabolic fuel. Cortisol is produced in the adrenal gland, which in turn is stimulated by ACTH from the pituitary. ACTH is necessary for the adrenal gland to develop properly as well as to function in producing cortisol.

In sheep, as in humans, the principal source of progesterone late in gestation is the placenta. Specialized regions of placental tissue called placentomes in the sheep contain the biochemical machinery necessary to convert cholesterol into pregnenolone, the first step in the synthesis of steroid hormones. Recall that steroid hormones are all closely related to each other, varying primarily in the number and type of side groups attached to a common core. They can be converted one to another in particular sequence by enzymes that remove or add these side groups. Along the pathways that these conversions create are a number of branching points, where a given steroid can potentially be converted into two others. Which of these conversion steps is followed, and indeed whether any conversion occurs at all, is determined by the presence and activity of the enzymes responsible for the alternative conversions. During gestation in the sheep, virtually all of the pregnenolone is converted directly into progesterone. The progesterone acts, as noted above, to maintain the myometrium in a quiescent state as the fetus develops and grows.

In the last few days of gestation the situation changes quite abruptly. The adrenal gland of the fetal lamb begins to secrete increasing amounts of cortisol. The cortisol has a number of important functions in addition to its primary role as a blood sugar regulator. It stimulates the final maturation of the fetal lungs and the production of specialized fluid necessary to keep the lungs from collapsing when they fill with air after birth. But it also stimulates the production of key enzymes in the placentomes, which direct the conversion of pregnenolone down an alternative pathway, away from progesterone production toward the eventual production of estradiol. The induction of these enzymes is akin to throwing a switch on a railroad track. The trains that used to arrive regularly at one station fail to show up and instead start to arrive at a different station. Over a few days the levels of progesterone in circulation drop to a small fraction of their former levels while levels of estradiol increase by more than thirty-fold. The sedating action of progesterone on the myometrium is abruptly terminated and replaced by the stimulating action of estradiol. At the same time that cortisol throws the switch on steroid production, it also stimulates the placental production of prostaglandins. The effectiveness of the prostaglandins in stimulating coordinated uterine contractions is in turn enhanced by the ebb of progesterone and the flood of estradiol. Within a few days of the first detectable rise in fetal cortisol, as a result of this hormonal cascade, parturition begins.

The leading event in this cascade is the production of cortisol by the fetal adrenal. In the fetuses of the sheep that feed on *Veratrum californicum* during gestation, the pituitary gland fails to develop the normal vascular connections to the hypothalamus. As a result, it fails to secrete ACTH in normal amounts and the fetal adrenal glands remain underdeveloped, incapable of producing the key event, an increase in cortisol, necessary to trigger spontaneous parturition. Hence the failure of labor to commence in these unfortunate animals. Surgical removal of the fetal pituitary or ablation of the fetal hypothalamus similarly delays the onset of parturition in sheep, while the exogenous administration of cortisol or ACTH will trigger premature delivery. Based on these observations the theory was advanced that the fetus, rather than the mother, initiates the process of parturition, probably coordinating this event with its own state of maturation.

The sheep model of parturition applies, with suitable minor modifications, to many mammals, but not to humans. In humans, all the same actors can be identified and their roles are recognizable, but the drama does not unfold according to the same script. Progesterone is certainly secreted in massive amounts throughout human pregnancy, at first by the corpus luteum of the mother's ovary, but after the first month or so primarily by the placenta. As in sheep, progesterone helps to maintain the pregnancy by sustaining the viability of the endometrium and sedating the myometrium. But progesterone levels in humans, unlike those in sheep, do not fall prior to the initiation of parturition; indeed, placental production is maintained to the very end. Progesterone levels in maternal blood fall only after the detachment and delivery of the placenta. Estradiol levels are high in humans just before labor begins, as they are in the sheep, but not as a result of an abrupt rise. Like progesterone levels, estradiol levels and the levels of another estrogen hormone, estriol, have been increasing throughout pregnancy. The rate of estradiol production increases toward the end of pregnancy, but over the course of weeks, not one or two days. The fetal adrenal gland produces cortisol in increasing amounts toward the end of pregnancy in humans, too, but there is no dramatic surge in levels, as there is in sheep, that might serve as a parturition trigger. Nor is it possible to initiate human labor abruptly by the exogenous administration of cortisol or ACTH. At least one of the routes by which cortisol stimulates parturition in sheep is missing in humans, the induction of a switch in placental steroid biosynthesis from progesterone production to estradiol production. The increase in estradiol production at the end of human pregnancy does not occur at the expense of placental progesterone production. Rather, the substrates for estradiol are produced by the fetal adrenal gland in increasing amounts, paralleling the increase in cortisol production. As a result, the dramatic, switchlike change in steroid hormone levels that occurs so elegantly in sheep over the course of a few days—cortisol surging, progesterone plummeting, estradiol soaring—is replaced by a much smoother progression—progesterone at a high plateau, cortisol and estradiol rising slowly and steadily. By monitoring steroid levels in the umbilical cord blood of a fetal lamb one can predict immanent parturition with confidence. An abrupt change in levels signals the onset of the cascade leading to birth within a day or two. The same prediction cannot be made from knowledge of fetal steroid levels in humans. The higher the cortisol levels, the more likely it is that labor will begin soon, but the probability of parturition onset doesn't change much from day to day.

One hormone that is familiar to many women who have been through labor plays an important role in parturition, but not in the early stages. Oxytocin is a protein hormone that is produced from the posterior pituitary gland of the mother. It is a powerful stimulant of the contraction of smooth muscle fibers such as the uterine myometrium. It also stimulates contraction of the smooth muscle fibers surrounding the milk ducts of the breast when a mother nurses her infant, causing milk to be squeezed into the collecting ducts around the areola of the nipple. This phenomenon, known as milk letdown, can even become a conditioned response to the baby's hungry cry. Oxytocin has also been implicated in smooth muscle contractions during orgasm. Spasmodic contractions of the myometrium during orgasm may produce negative pressure and help to draw semen into the uterus. During labor oxytocin release strengthens and accelerates the uterine contractions leading up to delivery. Synthetic versions of the hormone, such as pitocin, are often administered as intravenous drips to women in labor to help achieve this result. After delivery of the baby, oxytocin stimulates the contractions that help deliver the placenta and reduce the uterus to something approaching its nonpregnant size. (Encouraging the baby to nurse soon after birth is thought to aid in this process by stimulating additional oxytocin release.) But although oxytocin plays an important role in bringing parturition to a successful conclusion, it does not seem to play a role in initiating it. Oxytocin levels rise only after labor has begun, and exogenous administration is very ineffective at initiating labor. Oxytocin release and its associated effects appear to be more of a maternal response to the onset of labor than its cause.

The one piece that does seem to be the same in sheep and humans is the key role of prostaglandins. These hormones do rise in concentration at the very beginning of labor and continue to rise dramatically as labor progresses. They are powerful facilitators of myometrial contractions, and can be effectively used to induce labor. They also function to cause myometrial contractions during the shedding of the uterine lining in menstruation and are responsible for the associated pain and discomfort that many women experience at that time. They are even effective as abortifacients, especially when used in combination with a progesterone blocker such as RU486. Their role in human parturition is clearly central. But it is not clear what causes their levels to rise. Late in pregnancy the levels of arachidonic acid in the amniotic fluid, the substrate used for the formation of the critical prostaglandins (known as prostaglandins E2 and F2a), rise, suggesting an increase in the breakdown of fats, especially those known as glycerophospholipids. The source of arachidonic acid and the site of prostaglandin production appear to be the portion of the placenta derived from fetal tissues. As labor begins the levels of arachidonic acid first fall as prostaglandin production increases. But subsequently the levels of arachidonic acid rise and stabilize as glycerophospolipid metabolism increases to keep pace. Infections may trigger spontaneous labor by stimulating arachidonic acid and prostaglandin production as a part of the natural immune response. But the precise trigger for prostaglandin production in normal parturition has not been identified.

HUNGRY FOR FREEDOM

The puzzle can be summed up nicely in the words of one of the leading textbooks of obstetrics: "Teleologically it is satisfying, even tantalizing, to believe that the fetus, after key maturational events in vital fetal tissue and organs are initiated or completed, provides a signal to maternal tissues to commence parturition. Because of the attractiveness of this possibility, many investigators, including ourselves, have searched for a *fetal trigger* [*sic*] that would launch the procession of parturitional events. No such fetal signal has been found in human pregnancy and parturition."

One alternative is worth considering: Perhaps there isn't a trigger. The wording in the quotation above may reveal an unconscious assumption about the nature of the process. It implies that gestation provides an environment for the assembly and maturation of "vital fetal tissue and organs," and that once this process has reached fulfillment the time for birth has arrived. One can almost envision the fetal lamb, waiting impatiently for the last rivet to be placed and the last test to be run before pushing the button to initiate the release from its confinement. In fact it's only under these circumstances—when a relatively discrete "completion" of fetal development can be identified—that an abrupt fetal trigger makes functional sense. In adaptive terms, we might imagine that a pregnant ewe is at increasing risk of predation, for example, as her mobility decreases. The lamb's own survival probability might be greater on its own four feet, *if* it is capable of managing them. As soon as all systems are go—muscles, metabolism, central nervous system—the button is pushed, and within hours the newborn is keeping up with the flock on its own.

Other scenarios could be drawn to fit other organisms just as fancifully. The point is more sober. An abrupt fetal trigger for parturition makes sense only if there is an abrupt threshold of viability that can be associated confidently with some identifiable stage of fetal development. The more one considers the human case, the less likely it seems that such a condition holds. At some point the fetal head becomes so large that physical delivery becomes increasingly difficult. But even here there is no clear point at which one additional centimeter of circumference makes a crucial difference, and if there were it would depend as much on the mother's pelvic dimensions as on the fetus's cranium. That would be a different sort of threshold, a "negative" threshold at which viability might drop abruptly. Evidence for a "positive" threshold, at which viability abruptly rises, is very hard to identify. The crucial viability thresholds for many physiological parameters are passed sometime near the beginning of the third trimester; the threshold for brain development is still far off at its end. There is no point at which one can assert with confidence that there has been "enough" maturation (except by the standard of what is normal). One can only say that "more is better."

A critical threshold external to the fetus, which abruptly affected its survival probability, could also constitute a reason for a fetal parturition trigger, if the fetus could monitor the passing of that threshold in some way. If, for example, it was important for births to be synchronous in a large herd, of wildebeest for example, or to occur at a sharply defined season for reasons of food availability, and if there were some signal that would convey reliable information about these external factors to the fetus, then natural selection might favor the development of a fetal trigger. Conditions like these do not appear to fit the human case, however. If there were an innate limit on placental viability that the fetus could predict, that would qualify as such an external threshold. But as noted above, there is no evidence for such a limit.

Perhaps there is no reason to expect a fetal trigger for human parturition, at least not in the manner displayed in sheep. Still, the timing of parturition doesn't seem to be a random event. It occurs with enough regularity for doctors to predict due dates, but not with enough regularity for those predictions to be very impressive. The average length of gestation tends to be about forty weeks, but the variation around that average is considerable. Not many eyebrows are lifted when a birth is early or late by a matter of a few weeks. If there is simply a timer that goes off at nine months, it's a pretty inaccurate one. Nor has all the variation been removed by ultrasound assessments of fetal growth. Moreover, there are consistent patterns to some of the variation in gestational length. As noted above, undernourished mothers tend to have shorter gestations, as do mothers of twins and triplets, women who smoke, and women who live at high altitudes. Conversely, obese women and women with diabetes tend to have long gestations.

Postterm pregnancies, or those that continue for more than forty-two weeks, hold an interesting clue to the normal timing of birth. Several pathological conditions can lead to postterm pregnancies, such as maternal diabetes and fetal adrenal hypoplasia, or underdevelopment of the adrenal gland. At other times a specific cause is difficult to determine. Postterm pregnancies can lead to several difficulties that are not particularly surprising, such as hydramnios, or chafing and drying due to a paucity of amniotic fluid in the increasingly overcrowded womb, and an increased frequency of difficult delivery due to large fetal skeletal dimensions, both of the shoulders and of the head. But one common condition is rather surprising: fetal wasting, or loss of fat reserves. Although large in terms of their skeletal dimensions, postterm babies often lose weight, particularly fat reserves, at the end of their longer than expected gestations. Longitudinal fetal growth standards show this tendency clearly. Fetal growth in weight reaches a peak at about thirty-six weeks and then begins to slow rapidly. By forty weeks the rate of growth in weight is nearly zero, and by forty-four weeks it has become negative.

All of these observations lead to a new hypothesis for the timing of parturition: Birth occurs when the fetus starts to starve. At some point the metabolic requirements of the growing fetus begin to surpass the mother's ability to meet them. This is not a fixed developmental milestone for the fetus, though it may tend to occur around forty weeks of gestation in most pregnancies. The precise timing of this crossover of fetal demand and maternal supply will depend at least partly on the mother and the conditions that affect her metabolism. There are abundant indications that a crossover is inevitable. Early in gestation the energetic demands of the fetus are minimal. Weight gain in the first trimester is mostly a reflection of increasing fat deposition in the mother. In the second trimester the growing fetus becomes more of a metabolic load and maternal fat storage slows. By the third trimester maternal fat is mobilized to meet the increasing energetic demand of the fetus since the mother is unable to consume enough calories to keep pace. The mechanisms that a pregnant woman's body uses to provide energy to her ravenous third-trimester fetus are much the same as the mechanisms that a starvation victim's body employs to stay alive, prompting some physiologists to refer to the metabolic state of the pregnant woman as "accelerated starvation." The third trimester of pregnancy and the fetal brain growth and fat accumulation that have such an important influence on the fetus's probability of survival are only energetically possible because of the maternal reserves accumulated in the first trimester.

The physiological importance of first-trimester maternal fat storage has been underscored by research carried out in the Gambia under the auspices of the Medical Research Council of Britain and the Dunn Nutrition Laboratory of Cambridge University. A program was established in the Gambia in the 1960s to monitor maternal and child nutritional status and to develop interventions that would lower infant and child mortality rates. Baseline data on nutritional status and body composition indicated that the women in the study, residents of rural, subsistence farming communities, had very low fat reserves on average. Daily caloric consumption was below accepted minimum standards, often well below, while energy expenditure in subsistence activities was often high. The Gambian women were clearly just getting by energetically with little if any margin for error. Yet when they became pregnant they would put on fat during the first trimester, perhaps not as much as a pregnant woman in London, but a significant gain relative to their nonpregnant state. Careful monitoring of energy intake and expenditure did not reveal changes in consumption or activity patterns sufficient to account for the fat gain. It appeared as if the women were suddenly able to make a barely sufficient diet generate a caloric excess, stimulating hypotheses of some increase in "metabolic efficiency" that enabled them to get more calories out of a given food item than the average human.

Further research into possible energy-sparing mechanisms, however, revealed a different metabolic trick. During early pregnancy the average Gambian woman lowered her own basal metabolic rate, reducing her own energy requirement and freeing up energy to be stored as fat. Such a neat trick cannot be performed without cost, however. Cutting back on basal metabolism slows all sorts of physiological processes, compromising a host of functions from cognitive performance to thermoregulatory ability. One of the most significant costs may be reduced immune function. Pathological conditions such as hypothyroidism can produce many of the same effects.

A lowering of basal metabolic rate in early pregnancy is not a special trick of Gambian women, as it turns out. Some women in developed countries demonstrate it as well. It seems instead that this ability to cut metabolic corners in order to free up energy for fat storage is a general human ability that will be deployed when necessary. The very existence of such a mechanism underscores the high physiological priority a woman's body places on fat storage in the first trimester, even if it must be achieved at some significant cost to her own well-being. The reason, in turn, that first-trimester fat storage is so important is its crucial role in helping to meet fetal energy demands later in the pregnancy and so to stave off the point at which fetal demand exceeds maternal supply. Only by expending more calories than she consumes can a pregnant woman sustain her ravenous fetus in the final months of her pregnancy. But as the fetus continues to grow, so does its demand for energy. The mother cannot keep pace forever.

A particularly telling indication of the energetic constraints on human gestation is the form and function of the placenta itself. In the last chapter we noted that placentation in humans and other Old World primates represents one extreme on the mammalian scale of variation, an extreme known as haemochorial placentation. In this form of placenta all the vascular channels for maternal blood flow are lost, and the chorionic villi of the fetal circulatory system lie suspended in open cisterns of maternal blood, the intervillous space. This peculiar arrangement has several distinct drawbacks. High concentration gradients must be maintained to achieve adequate transfer of key nutrients and gases, and high blood pressure must be maintained to sustain adequate circulation. The spiral arteries that lead to the intervillous space help to provide adequate pressure by acting as pressure valves. Only when the head pressure of the blood exceeds the resistance of the arterial coils is the blood released in a jet.

In the 1960s Elizabeth Ramsey and her colleagues provided stunning documentation of the dynamics of placental blood flow. By injecting special dyes into the uterine artery of pregnant subjects, Ramsey was able to produce motion picture x-rays of maternal blood pulsing into the intervillous space like geysers from the openings of the spiral arteries, being forced farther through the space by succeeding bursts, and finally draining back to the maternal circulation through venous openings in the floor of the maternal side of the space. But even with this elegantly engineered architecture, maintaining an adequate circulatory flow is a difficult feat. Maternal blood pressure has to remain high for the system to work. Similarly, high levels of key nutrients and gases must be maintained in the maternal blood to sustain adequate transfer to the fetus. Special mechanisms are employed to aid in the transfer of the most important molecules, such as glucose and oxygen. Glucose is the basic energy substrate for fetal metabolism, and the only such substrate that can be transferred across the placenta in sufficient quantity. Oxygen is necessary to "burn" that fuel and produce utilizable metabolic

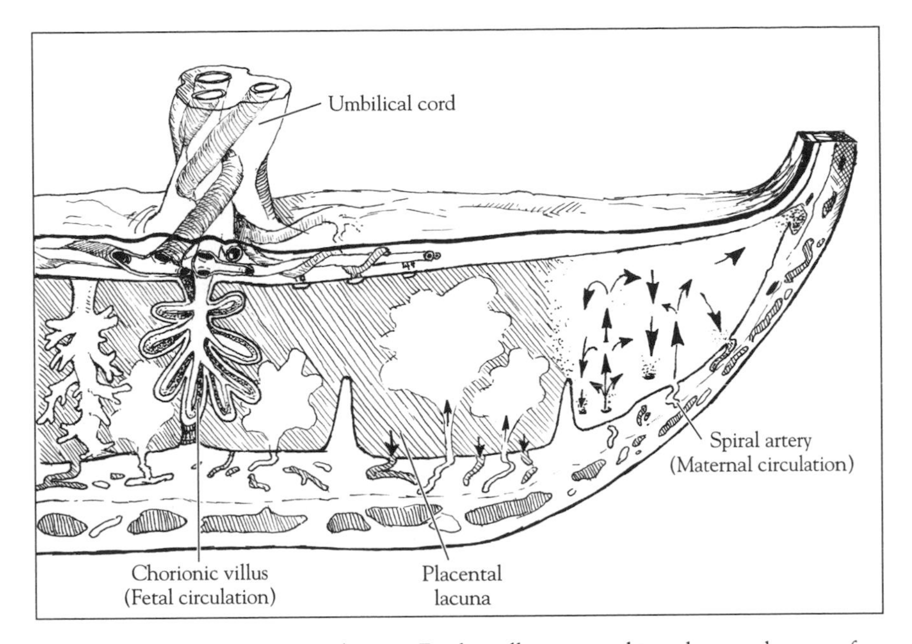

Blood circulation in the human placenta. Fetal capillaries extend into the open lacunae of the placenta as highly branched chorionic villi. Maternal blood is released in jets from spiral arteries to bathe the villi before draining into the maternal venous system.

energy. An active transport enzyme helps shuttle glucose from the maternal to the fetal circulation. Oxygen uptake by the fetal circulation is aided by a fetal form of hemoglobin that has a higher binding affinity for oxygen than does maternal hemoglobin. But their levels in the mother's blood still primarily determine transfer of both of these substances. Fetal growth is slower at high altitude than at low altitude, principally as a result of the lower oxygen levels in the maternal circulation. Special adjustments in maternal metabolism, some stimulated by placental lactogen, a special metabolic hormone secreted into the maternal circulation by the placenta, help to maintain blood glucose levels at the high end of the normal range in pregnant women. Increases in blood volume and vascular tone help to keep maternal blood pressure high as well so that blood flow through the intervillous space is sufficient to sustain the transfer of these nutrients and gases to the fetus.

The necessary adjustments to maternal blood pressure and glucose levels can have dangerous consequences, however. The typical tendency to hypertension in pregnancy, if pushed too far, can result in a dangerous and even fatal condition known as preeclampsia, while the mechanisms leading to elevated glucose levels can result in gestational diabetes. The inefficiencies and risks inherent in this form of placentation are offset, however, by one principal advantage: a tremendous increase in the surface area over which nutrient and gas transfer can occur. Even with the rate-enhancing effects of fetal hemoglobin and active glucose transfer, not enough of these and other crucial nutrients could cross the placenta in a given unit of time if the area available for transfer were limited to the adjacent surfaces of juxtaposed capillaries. The maximal rate of glucose transfer per gram of placental tissue, for example, is ten times higher in a human than in a sheep. At the beginning of a pregnancy this enhanced capacity of the placenta for nutrient and gas transfer isn't crucial. It's at the end of the pregnancy, when every effort is being made to push enough calories across from the mother to the fetus, that the need for enhanced transfer capacity becomes acute.

The optimal timing of birth, then, may be determined not by an abrupt viability threshold tied to fetal maturation, but by the ability of the mother to continue to meet the energetic demand of the fetus. Natural selection has adapted placental morphology to provide for an enhanced rate of nutrient transfer, but at some point maternal supply and fetal demand must cross over. A fetus that is "forced" by some circumstance to stay in the womb beyond this crossover point must turn to the digestion of its own fat reserves to sustain itself, hence the fetal wasting in postterm babies. But the timing of the crossover will not be absolute. It won't happen magically at nine months. The timing will depend at least partly on the mother and her energetic status. A mother with abundant energy availability in her blood, such as a diabetic mother or an obese mother, will be able to keep up with fetal demand for longer. A mother with low energy availability, because she is herself undernourished or was unable to accumulate sufficient energy reserves early in the pregnancy, will reach the crossover point sooner.

Variation in fetal demand will matter as well. Twins and triplets will result in an earlier crossover point than singletons. Anencephalic infants, like the one mentioned earlier, may generate particularly low energy demands since the most ravenous part of their anatomy, the neocortex of the brain, is missing. Differences in oxygen availability to the fetus may also affect the crossover point, although such differences are not common. High altitude is associated with shorter gestation length, at least among mothers who are not native to high altitudes. Smoking, too, shortens gestation, at least in part because of the increase in carbon monoxide in the maternal blood and the associated reduction in oxygen available to the fetus.

Understanding of these facts also dissolves the paradox of underweight babies. It would do no good to prolong gestation in an undernourished mother in whom the crossover point is reached sooner than average and at a lower fetal weight. Continuing the pregnancy would not allow the infant to accumulate more fat, but only force it to begin using whatever reserves it has. The interaction between fetal growth rate and the attainment of the metabolic crossover point may in fact serve to reduce the variance in gestation length somewhat. Undernourished women might have a lower crossover point but might take longer to reach it owing to slower fetal growth. Well-nourished women might have a higher crossover point but reach it more quickly. It's interesting to note that when pregnant women in the Gambia were given caloric supplements to their diets, average birthweights increased only a very modest amount. But the frequency of very low birthweight babies, those under 2,500 grams, and the associated frequency of preterm births decreased significantly. Conversely, during the famine period of the Nazi occupation of the Netherlands, when maternal caloric intakes decreased significantly, the frequency of preterm births increased.

The critical parameter determining the timing of parturition, then, may not be a threshold of fetal maturation, or even an absolute level of fetal energy demand, but a balance between that demand and the mother's ability to meet it. If we revisit the physiology of parturition from this perspective, the conundrum of the missing fetal trigger may also seem less puzzling. The increasing fetal cortisol production over the last weeks of pregnancy may reflect the increasing difficulty the fetus is experiencing in meeting its own energetic requirements. The rising cortisol levels stimulate the final preparation of the lungs for the birth that seems more and more imminent. As the adrenal gland of the fetus increases its output of cortisol, it also increases the output of substrates for estradiol production by the placenta, increasing the contractile tone of the myometrium. At some point the fetal demand for energy, reflected in the fetus's increasing cortisol levels, begins to stimulate the breakdown of fats. If this metabolic process extends to the placenta, then an increase in arachidonic acid production could result, eventually leading to the production of prostaglandins. Rather than a triggering event, the onset of parturition may be more like the collapse of a sand pile that has been pushed higher and higher. With each added grain, each increase in fetal energy demand, the event becomes more likely, the maintenance of the status quo more fragile. The rate at which the consequences of fetal energetic stress accumulate may differ substantially in different pregnancies under different ecological circumstances. Rather than being a fixed developmental trigger for initiating parturition, this mechanism would allow for a more flexible onset of labor still directed toward optimizing the survival chances of the offspring.

As noted in the last chapter, the genetic interests of a mother and her fetus do not coincide. Conflict can be predicted over issues such as the optimal allocation of maternal energy to the mother or to the fetus. Some conflict might be predicted over the optimal timing of birth as well. Under the metabolic crossover hypothesis, however, such conflict might be minimal. It is in the interests of both mother and fetus to enhance the baby's postpartum survival probabilities as much as possible. It is not in the interests of either to prolong gestation into the region of fetal wasting. Fetal demand may push the mother to the brink of her metabolic capability, but pushing further may not be advantageous if it compromises the continued investment the infant will require from the mother after birth. In any event, there is nothing in the mechanisms controlling human parturition to undermine the idea that the fetus rather than the mother sets the process in train. If we imagine that the fetus is in total control of initiating parturition and that the mother has no say, the crossover hypothesis still applies. Gestation should end when the fetus starts to starve.

If birth is a consequence of fetal metabolic demand outstripping maternal supply, one might wonder if it really solves the problem. After birth the infant is, at least under "natural" conditions, still dependent on the mother's physiological ability to meet its nutritional needs, albeit through the nipple rather than across the placenta. The infant's needs are even increased by its new, extrauterine existence. It must maintain its own body heat, breathe for itself, and take on numerous metabolic tasks previously performed at least in part by its mother's physiology. And of course it is still growing at a dramatic rate. Even as the infant's metabolic requirements increase, energy transfer from the mother requires extra steps after birth. Rather than being able to pass nutrients directly from her own bloodstream to that of the fetus, the mother must now synthesize milk for the baby to ingest and digest. A common estimate is that this more indirect provisioning carries a 10 percent tax: that it takes 100 calories on the mother's part for every 90 calories delivered to the infant, the rest being used to pay the costs of milk synthesis. An additional unestimated "tax" occurs in the infant to cover the metabolic work of digestion, though this may be better thought of as a part of its increased metabolic requirements.

But how, if the mother has reached a point at which she is no longer able to meet the metabolic demands of the fetus in the womb, will she be able to meet the even greater demands of the infant out of the womb through a process of nutrient transfer that is more costly? The answer is that the production of milk allows the mother to increase dramatically the transfer of fat as well as simple sugars to the infant. Fat does not cross the placenta in appreciable amounts, so that the mother must provide calories to the fetus primarily in the form of sugars like glucose. Producing these sugars from her own fat reserves in adequate amounts becomes increasingly difficult, and the accumulation of byproducts of this process in her own circulation can rise to dangerous levels near the end of pregnancy. But in producing milk she may now pass fats directly to her offspring, providing a much more calorie-dense food from her nipple than she could ever produce in her own blood or deliver across the placenta. If fetal starvation at the end of gestation is the problem, milk is the answer and birth merely the necessary transition.

The culprit in the drama of human birth is clearly the fetal brain. It is the precocious enlargement of this organ that causes the fetal cranium to swell hugely at an early stage and hence to produce a selective pressure on female pelvic size. It is also the fetal brain that generates the very high metabolic demand at the end of pregnancy. The trend in this direction started well before the emergence of the human lineage. An enlarged fetal cortex is a characteristic of Old World primates in general, as is the development of a haemochorial placenta with its special adaptations for raising the ceiling on nutrient transfer. But in humans, brain size becomes enormous even by primate standards. As the story of the Garden of Eden suggests, one price of this burgeoning wisdom may be difficulty and pain in childbirth, a birth that is forestalled as long as possible and that requires the assistance of other humans to succeed.

THE ELIXIR OF LIFE

IT IS SOMETIMES SAID that mother's milk is nature's perfect food. If it isn't perfect, it is certainly impressive. It can be synthesized efficiently from the mother's primary energy reserves. It provides essential nutrition in an easily digestible form, both macronutrients (calories, proteins, fats) and micronutrients (vitamins and minerals). Ingestion of breast milk does not expose the infant to external pathogens the way ingestion of other foods or water may. But more that that, breast milk also includes immunologically active substances that help to protect the infant's gastrointestinal tract from pathogens that it might ingest at other times. Some of these substances are nonspecific, meaning they confer limited protection against a wide variety of pathogens. An example is lactoferin, an iron-binding protein that makes it difficult for bacteria in the gastrointestinal tract to get hold of the iron that their own metabolic processes require. Other substances are more specific and more potent, such as the antibodies a mother secretes directly into her milk. These cornerstone molecules of the vertebrate immune system recognize and bind to specific pathogens, activating mechanisms for their elimination. The immune system "remembers" pathogens it has been exposed to in the past and can rapidly mount an effective defense when reexposed. An infant's immune system is underdeveloped and does not have the reservoir of acquired "experience" that the mother's immune system has. However, since mothers are generally exposed to the same pathogens as their infants, the efficiency of the mother's immune system can continue to provide protection to the infant through breast milk.

The immunological benefit of breast milk is often made clear when it is removed. Weaning, especially if it is abrupt and occurs under conditions of high pathogen exposure, often results in "weanling diarrhea." This phenomenon was first observed in India by public health researchers in the 1960s. A cultural preference for weaning children at the age of about one year was associated with a dramatic increase in diarrhea and other gastrointestinal diseases and also with a drop in weight gain over the ensuing six months or more. Even when the transition from breast milk is gradual and occurs under conditions of low pathogen exposure, the incidence of diarrhea and other infections can increase as a consequence. A recent analysis of national data from the United States found that the frequency of gastrointestinal and respiratory diseases during the first six months of infancy was inversely related to the percentage of breast milk in the baby's diet: the more breast milk, the fewer infections. The substitution of powdered formula for breast milk can be particularly hazardous in areas with contaminated water supplies. Under such conditions the infant is simultaneously deprived of its best immunological defenses and exposed to concentrated pathogens in the very food that is being substituted for mother's milk.

SUPPLY AND DEMAND

Preparation for milk production, like other aspects of reproductive maturation, begins at puberty. Milk is produced by glandular tissue lining tiny sacklike structures known as alveoli, which are themselves clustered in bunches, or lobules, throughout the breast. The lobules are suspended in a matrix of connective and adipose tissue. A system of branching ducts spreads out from the nipple to connect all the lobules. The entire system is known as the lobular-alveolar duct system. It is latent in the undeveloped breast tissue of both boys and girls prior to puberty. At puberty the activation of the gonads leads to very different profiles of steroid hormones in males and females. The steroids in turn are responsible for the development of most secondary sexual characteristics, including breast development in girls, for which production of estradiol by the ovary is the most important signal. Estradiol stimulates cell division and tissue proliferation in the lobular-alveolar duct system and also promotes the accumulation of fat by breast adipose tissue. Breast tissue, particularly in the lobular-alveolar duct system, is richly endowed with estrogen receptors so that a significant response is initiated early in pubertal development. Estradiol stimulates cell division in many of the tissues it affects, including the endometrial lining of the uterus, and also in the very cells of the developing ovarian follicle that are responsible for producing estradiol. There is even evidence that estradiol helps to boost the high rate of cell proliferation on which immune responses are based, contributing to the fact that females generally outperform males on measures of immune function, and possibly contributing to generally lower mortality rates in women compared to men. This basic action of estradiol as a mitotic (cell-division) stimulant also leads to its role as a risk factor for many cancers.

Incipient breast development is usually one of the first external signs of early puberty in girls, indicating that estradiol production is beginning. The maturation of the lobulo-alveolar duct system is still incomplete, however, at the end of puberty. In particular, the glandular tissue of the alveoli remains undifferentiated. Once differentiated into final functional form, glandular epithelial cells lose the capacity for further mitotic division. They can only be renewed by replacement from a reservoir of undifferentiated tissue. The metabolic activity of mature glandular tissue is also generally quite high, since it is the role of such tissue to synthesize, package, and secrete particular molecular products. Hence glandular maturation usually doesn't occur until the secretory product is needed. At the end of puberty the duct system of the breast has been developed and the architecture of the lobular clusters of alveoli has been established, but the final maturation and differentiation of the glandular epithelium itself has been postponed.

Pregnancy provides the signals for final maturation of the lobulo-alveolar duct system. Sustained by the production of hCG by the embryo, the corpus luteum produces increasing amounts of estradiol as well as progesterone during the first weeks of pregnancy. Cell division within the lobulo-alveolar duct system is stimulated again along with additional fat accumulation. Changes in breast volume and tenderness are noticeable to a woman early in her pregnancy, long before abdominal bulges are apparent. Other hormonal signals contribute to establishing the metabolic machinery for milk production. Placental lactogen and cortisol help to stimulate the development of the enzyme systems necessary for the production of milk nutrients and prepare for the preferential diversion of maternal energy stores to this end. Once milk production begins, the glandular tissue of the breast will be particularly efficient at absorbing fats, sugars, and amino acids from the mother's bloodstream to use in milk production, while breast adipose tissue will be particularly efficient at absorbing glucose to store as fat for use as a primary substrate of milk production.

But while the end of pregnancy prepares all the machinery for milk production, the assembly line is not set in motion. The high levels of progesterone and estradiol circulating throughout pregnancy keep milk production in check. As soon as the placenta is delivered after birth, however, the source of the high steroid levels of pregnancy is removed and the cascading processes of milk production begin. Immediately after birth a nursing baby will draw no milk from its mother's breast, although it will obtain a thicker, waxy substance known as colostrum that is rich in antibodies and other immunologically active substances. Within hours, however, milk will reach the collecting ducts around the areola, where it is available to the nursing infant.

As well-developed as the cheek muscles of a newborn are, drawing milk forward all the way from the alveoli by vacuum pressure alone would be a difficult if not impossible task. Milk is in fact squeezed toward the nipple by the contractions of the smooth muscle fibers lining the alveoli and the milk ducts. This contraction is stimulated by oxytocin, the same hormone that promotes contractions of the uterus during labor. Oxytocin is released from the mother's posterior pituitary gland whenever the baby nurses. A direct neural circuit can be traced from the tactile receptors of the nipple to the posterior pituitary, so that the very act of suckling at one end of the lobulo-alveolar duct system is coupled to a squeezing action at the other end. Ordinarily, the squeezing action does not occur unless there is a mouth there to receive the milk. However, direct neural circuits such as the nipple-to-pituitary circuit are highly susceptible to classical Pavlovian conditioning. Hence the automatic "milk let-down" that many nursing mothers develop in response to their infant's hungry cry. (Since orgasm can also involve oxytocin release to help stimulate the smooth muscle contractions of the lower reproductive tract, milk let-down or even minor "spurting" can also accompany sexual intercourse in nursing mothers.)

Nursing also triggers another hormonal response in the mother that maintains the production of milk. The release of prolactin, a protein hormone, from the anterior pituitary gland is caused by stimulation of the nipple, the withdrawal of milk from the alveoli, and the emptying of the milk ducts. If the nerve supply to the nipple is removed in an experimental animal, release of oxytocin in response to suckling is blocked. Prolactin, however, is still produced in response to milk withdrawal and milk production continues. It is not clear how the signal that leads to the release of prolactin is transmitted from the breast to the pituitary in such cases. Perhaps some change in the metabolic activity of the glandular epithelium of the alveoli is coupled to the production of hormonal signals or changes in the concentrations of metabolic byproducts that register in the hypothalamus. Various hypothalamic hormones are known to affect prolactin, some stimulating and some inhibiting its release. Normally, however, nervous signals carry the message from the nipple to the hypothalamus. Within minutes after a baby begins to nurse, increases in prolactin concentrations are often detectable in the mother's blood.

Prolactin, as its name suggests, is the primary hormone responsible for sustaining the production of milk. It stimulates the glandular cells of the lobulo-alveolar duct system to transcribe the genes for assembling the characteristic proteins in milk, including casein and lactoalbumin, as well as to generate the other important constituents of milk, such as lactose and fatty acids, via enzymatic pathways. Two other hormones, cortisol and insulin, that are important in this process work synergistically with prolactin. These two are potent regulators of metabolism, helping to control the distribution of available energy among the body's tissues. Ordinarily cortisol and insulin have broadly antagonistic effects. Cortisol stimulates the release of energy into the bloodstream in the form of glucose and free fatty acids, the glucose coming primarily from liver glycogen and the fatty acids from the breakdown of fat in adipose reserves. Insulin encourages the removal of energy from the bloodstream, stimulating uptake of glucose by muscle and liver cells for storage as glycogen and by adipose cells for incorporation in fat. At the same time insulin suppresses the release of glucose and fatty acids from storage tissues. One of the effects of prolactin, however, is to increase the density of insulin receptors on the milk-producing cells of the breast. Low levels of insulin thus end up promoting milk production by supporting the uptake of glucose and fatty acids by the milk-producing cells without shutting off the processes that keep those molecules flowing out of their storage depots. High levels of insulin, however, when introduced experimentally into lactating animals, reduce milk production by shunting available energy away from the breast toward storage tissues. Cortisol supports the action of prolactin both by keeping the supply of energy substrates flowing and by helping to regulate the genes governing the production of the milk-producing enzymes. As the infant grows and its dependence on breast milk decreases, cortisol levels in the nursing mother tend to fall and insulin levels to rise.

It is prolactin, however, that provides the primary stimulus for milk synthesis, and prolactin release in turn is linked to the withdrawal of milk from the breast. The demand for milk by the infant is thus naturally part of the feedback system that sustains the supply. If a woman does not nurse her baby after birth, milk production will stop fairly soon. As a baby matures and nurses less, less milk is produced. When a baby is finally weaned, milk production ceases.

Some things can disrupt this elegant feedback system, however. The longer the interval between birth and the establishment of regular nursing by the infant, the harder it is to start milk production. Similarly, interruptions of regular nursing for more than a day or two, by maternal or infant illness for example, can begin to shut down milk production and make it difficult to restore it. Even a condition as common as cracked nipples, by making nursing painful, can threaten to undermine the feedback loop of supply and demand that keeps milk production going. Fortunately, there are organizations in many countries, such as the La Leche League, that provide information and support for women who may be experiencing difficulty in establishing or sustaining a regular pattern of breastfeeding.

High levels of ovarian steroids, particularly estradiol, can also inhibit milk production. Estradiol counteracts the effects of cortisol and suppresses the biochemical machinery of milk production. It also inhibits the flow of metabolic substrates to the glandular tissue of the breast by stimulating the uptake of these molecules by adipose tissue elsewhere in the body. Ordinarily this antagonistic action of estradiol helps to couple the cessation of milk production to the natural resumption of ovarian cycles, a phenomenon that we will consider in detail below. But exogenous steroids can have the same effect. Oral contraceptives, for example, can inhibit milk production if taken during lactation. For this reason the timing of the introduction of steroid contraception after birth in populations where infants depend heavily on the nutritional and immunological benefits of breast milk is a matter of serious public health concern.

GOING WITH THE FLOW

The fact that milk production is matched so closely to demand makes physiological sense when we remember that it represents a considerable metabolic load on the mother. Lactation may have an advantage over placental transfer in terms of the caloric density of the nutrition provided and the maximal rate of transfer attainable, but that doesn't mean that the absolute cost of remaining the baby's primary source of nutrition declines after birth. Indeed, the metabolic cost of supporting the infant continues to increase until foods other than breast milk begin to constitute a significant portion of its diet. Estimates of the caloric cost of lactation vary, but generally fall in the vicinity of 500 kilocalories a day, or an additional 15–25 percent of the mother's daily energy budget, and even more if she is living on a marginal diet. ("Kilocalorie" is the technical term that is often shortened to "calorie" by dieters and food manufacturers.) While underproduction of milk might threaten the baby's survival, overproduction of milk could constitute an unnecessary and even dangerous burden on the mother. The feedback between infant demand and maternal production helps to assure that the mother produces enough milk, but not too much nor for too long.

The cost of milk production can be met in a number of ways. Some of the cost can be met by increasing food consumption. Some can be met by the mobilization of stored fat reserves, continuing the trend begun in the third trimester of pregnancy. Additional energy may also be made available for lactation by the reduction of expenditures on physical activity. In many populations, however, particularly those in rural areas of the developing world, these options may all be limited. Increases in consumption may be difficult when food is in short supply. Fat reserves are often limited for the same reason. As for reduction in physical activity, field studies of many rural populations in the developing world have demonstrated remarkably little difference in activity patterns between lactating women and those neither pregnant nor lactating. Instead, researchers are often struck by the rapidity with which new mothers resume their accustomed share of subsistence and household work. In Nepal, for example, Catherine Panter-Brick has found that lactating Tamang women work just as hard as their nonpregnant, nonlactating peers during the monsoon season, when the heaviest labor demands of the agricultural cycle fall.

Physical activity is not the only metabolic corner that can be cut, however, to free up extra energy for lactation. Reductions in basal metabolic rate, particularly during the early postpartum months before substantial supplementary food is introduced to the infant's diet, have been demonstrated among women in the Gambia and elsewhere. But a strategy as drastic as reducing one's metabolism is not infinitely elastic. At some point reductions in metabolic rate would seriously threaten a mother's survival. When energy constraints become so severe that such cost-cutting measures no longer suffice, milk production may suffer. In the Gambia, seasonal food shortages have been associated with a dramatic pattern of seasonal weight loss as daily caloric consumption falls to about 1,000 kilocalories a day. Under these extreme circumstances metabolic savings cannot make up for the cost of milk production. Even maintaining a reduced metabolism requires rapid withdrawal of stored energy reserves. As a result, milk output among lactating mothers was observed to decline in parallel with body weight during these periods of dramatic energy shortage.

Yet milk production is afforded a very high metabolic priority and buffered to the extent possible as long as infant demand is there. This fact was underscored by a later stage of the research in the Gambia when a program of nutritional supplementation of pregnant and lactating women was undertaken in an effort to reduce infant and child mortality and morbidity and to improve infant growth rates. The theory behind the intervention was straightforward: since the mother provides virtually all the nutrition to a growing fetus and a breastfeeding baby, the best way to improve the offspring's nutritional status is to improve its mother's. The assumption is that the mother will "share" (physiologically) her supplemental nutrition with her baby. The supplementation program was undertaken very carefully with thorough follow-up. Supplements were produced locally from traditional food sources. Observations were made to document the net effect that the supplements had on the mothers' diets. Although women did tend to reduce their consumption of other foods when provided with a supplement, the program managed to achieve an impressive average increase in maternal energy intake of around 725 kilocalories a day, a 45 percent increase over the unsupplemented average intake. The supplement also increased protein intake by 50–100 percent.

The effect of this intervention on the outcome variables of interest, however, was minimal. Average birthweights increased by only some fifty grams. Average milk output did not change at all! Nor did the composition of the milk change dramatically. A small increase in protein content (7 percent) was observed in the milk of the supplemented women, but the energy content remained the same. A slight increase in fat content was offset by a slight decrease in sugar content. Although this result was not anticipated in the design of the intervention, it is consistent with the notion that milk production is highly buffered from variations in maternal nutrition. When maternal energy intake is low, maternal metabolism is reduced in order to maintain milk production. When maternal energy intake recovers, in this case as a result of the supplementation program, the mother's metabolism is restored while milk production remains unchanged. Only when caloric intake approaches starvation level is milk production significantly curtailed. Across a broad range of caloric intakes above this minimum, however, milk production is maintained at a level matched to demand.

Milk production has proven very robust and hard to perturb among women in the developed world as well. Exercise, for example, does not necessarily suppress milk production. In California, neither a study comparing lactating women who exercised voluntarily with nonexercising lactating mothers, nor a study that randomly assigned nursing mothers to exercise and control groups, found any effect of the exercise on milk production or composition. Increases in maternal energy expenditure were partially met by increases in energy intake, but not entirely. Additional savings may have been achieved by adjustments in the mother's basal metabolic rate.

Two aspects of milk composition appear to be more susceptible to dietary influence: fat content and water-soluble vitamin concentration. If the fat content of a nursing woman's diet is increased substantially to the point of constituting 50–70 percent of her caloric intake, the lipid and fatty acid concentration of the milk she produces also increases. This implicates the mobilization of fat stores as a limiting feature of milk production. When fat must be mobilized to meet the mother's own survival needs, the rate of milk production may decline. When excess fat is available in the bloodstream without the mobilization of reserves, the fat content of the milk may rise. The increase in the fat content of the Gambian women receiving dietary supplements may reflect an increase in their ability to accumulate breast fat for milk production. Supplementing the mother's diet with these specific nutrients can also increase the concentration of water-soluble vitamins in the milk, such as most of the B vitamins and vitamin C. These vitamins are not susceptible to maternal storage; hence their appearance in the milk in concentrations that directly reflect maternal blood levels is not surprising.

Researchers at Cornell University have used the term homeorhesis to refer to the metabolic priority given to lactation in the partitioning of maternal resources. In contrast to homeostasis, referring to the maintenance of a static physiological condition, homeorhesis refers to the maintenance of a flow of resources to a metabolic process. As is the case in most examples of homeostatic mechanisms, homeorhesis is often accomplished by the hormonal regulation of metabolism. Prolactin is an excellent example of a hormone that acts to adjust the total flow of metabolic resources to milk production. Not only is the enzymatic machinery of milk production activated, but the mobilization of fat reserves is potentiated and the storage of fat other than in the breast is inhibited. The flow of energy toward milk production and away from other tissues is aided by the increase of insulin receptors in the breast and their decrease in other tissues. The mechanisms that normally divert energy in other directions are suppressed.

A lactating woman who is subjected to exercise stress on a treadmill shows very little increase in the hormones that maintain glucose homeostasis, such as cortisol and epinephrine. In a nonlactating woman these hormones rise in response to exercise to promote the release of sugar from the liver and its delivery to the muscles. In a lactating woman the same hormones rise very little if at all. Rather than rising or remaining steady, her blood glucose levels fall as a result of the exercise. Nor can a lactating woman be easily "forced" to divert metabolic resources in other directions. Resting metabolism usually rises after a meal, in response to a release of the adrenal hormone norepinephrine. In a lactating woman resting metabolic rate does not change after a meal, or in response to an infusion of exogenous norepinephrine. It is as if the metabolic conduit leading to milk production is enlarged and competing conduits are reduced or closed off by prolactin. Or to use a different analogy, imagine that human metabolism is like a billiard table, with billiard balls representing quanta of available energy and the pockets around the table the possible destinations for available energy: storage, protein anabolism, muscle utilization, milk production, and so forth. Prolactin tips the table toward the pocket representing milk production, differentially directing available energy to that end and making it more difficult to divert energy to other purposes.

Lactation, then, is best understood as an extension of the mother's direct physiological investment in her offspring after birth. The cost of this continued investment is high, anywhere from 15 percent to 50 percent of a mother's total energy budget, depending on her circumstances. Meeting this burden depends on the efficient mobilization of fat stores for the production of milk. The process of milk production is afforded a high metabolic priority and is broadly buffered against variations in the mother's energetic state by the homeorhetic action of prolactin. The continued production of prolactin, and hence of milk, is coupled to the infant's consumption. As the demand for milk declines and eventually ceases altogether, so does the stimulus for its production.

METABOLIZING FOR TWO

Both gestation and unsupplemented lactation represent very demanding energetic states for a woman. She is at these times essentially metabolizing for two as she attempts to meet both her own nutritional needs and those of her offspring. Meeting these simultaneous demands often entails a significant cost to the mother, in terms of either reduced metabolism or depleted energy reserves or both. Keeping a sufficient separation between successive intervals of lactation and gestation may be crucial in maintaining a woman's long-term energy balance and preventing a downward spiral of maternal condition. Such a downward spiral resulting from closely spaced periods of lactation and gestation has been termed maternal depletion. As a phenomenon, maternal depletion was first described among the natives of highland New Guinea, but the term was soon broadened into the notion of a "maternal depletion syndrome" of wide distribution in the developing world. The syndrome is particularly likely to occur where high female workloads and poor nutritional intake compound the energetic drain of closely spaced births. Under these conditions female fat reserves can be shown to fall progressively with increasing parity (number of births). In addition to undermining the mother's own health, depleted maternal reserves can jeopardize the health and survival chances of her offspring. Depleted reserves can reduce a mother's ability to invest in her offspring during gestation, leading to low birth weight, early parturition, and increased risk of infant mortality. Low reserves can also, as noted above, result in reduced milk production, thus lowering the postnatal investment a mother can provide to an infant who may already suffer from low birthweight. It is easy to imagine how serious the negative consequences of maternal depletion can be, and therefore how important it is for a woman's overall evolutionary fitness to maintain a sufficient separation of periods of metabolizing for two.

Data from around the globe have clearly demonstrated some of the costs. In particular, among populations without widespread use of contraception, the shorter the interval separating births, the greater the risk of mortality for the offspring. Intervals of less than two years between births are particularly dangerous. Interestingly, the increased risk occurs for both the first- and second-born offspring in a sequence. One might imagine that the early conception of a younger sibling might lead to early weaning of the older child, exposing it to increased risk of infection as well as undermining its nutritional status. But in a number of studies, an increased risk to the younger child has also been observed. In a survey of data from twenty-three countries, for example, it was found that an interbirth interval of less than two years was associated with an increase of over 50 percent in the risk of the younger child's dying in thirteen of the countries. In nine of those countries there was also evidence that a previous history of closely spaced births (two or more births in the two to six years before the birth of the index child) increased the risk of death to the index child even further. In Nepal this effect of closely spaced births was found to be independent of the nurs-

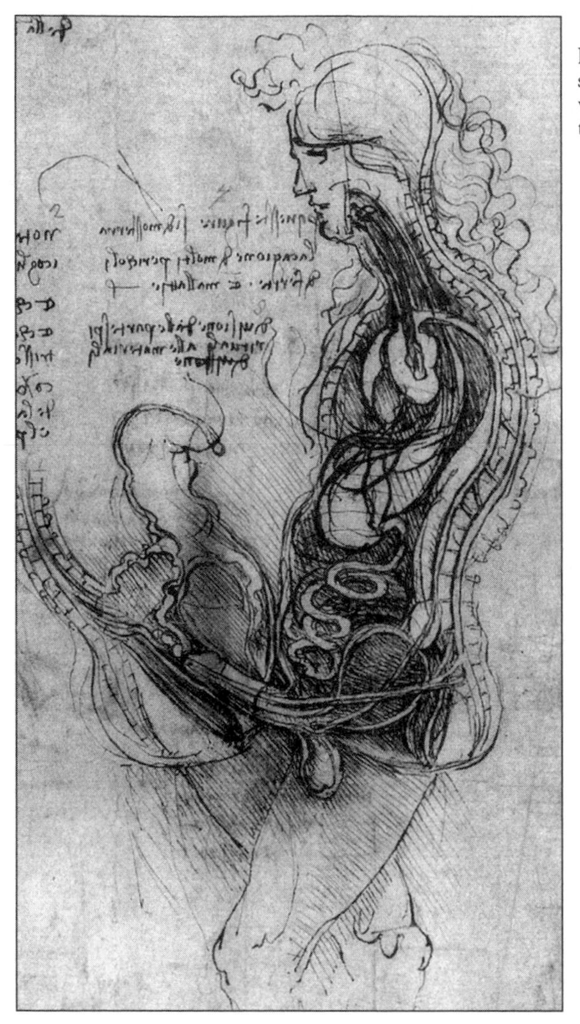

Leonardo Da Vinci's drawing showing the imaginary "milk vein" connecting the uterus to the nipple.

ing status of the older sibling: It did not appear to be competition for the mother's breast that was responsible for the increased risk to the younger child. Rather it seems that closely spaced births have a cumulative negative effect on the mother's ability to invest physiologically in her current and subsequent offspring. Nor is this effect limited to populations living under marginal nutritional circumstances. Even in contemporary Utah a negative effect of closely spaced births on birthweight and the risk of prematurity has been documented.

If metabolizing for two is a significant burden, metabolizing for three is nearly impossible, especially when the mother's energy budget is constrained by external circumstances. It should come as no surprise, then, that natural selection has provided physiological mechanisms to reduce the likelihood of overlapping gestation and lactation. The fact that these two reproductive states appear to be mutually exclusive, that a nursing woman is unlikely to be menstruating much less to get pregnant, was well known to Aristotle, Galen, and medieval physicians. One of Leonardo da Vinci's famous anatomical drawings clearly shows the "milk vein" connecting the uterus and nipple. The depiction of this imaginary bit of anatomy reflected the widely held medieval notion that the breast and womb represent alternative destinations for the flow of vital fluids. Vital fluids can go to the uterus for the nourishment of a fetus or to be discharged as menstrual blood, or they can go to the breast to produce milk, but not both. The experience of wet-nursing, common in Europe from antiquity into the eighteenth century, underscored the contrast between the natural mother, who might begin menstruating again soon after giving her child over to a wet-nurse, and the wet-nurse herself, who might remain amenorrheic for years while nursing a succession of infants.

A MALTHUSIAN LEGACY

Yet as obvious as it may have been to both physicians and laity in the past, the notion that lactation somehow suppressed a woman's reproductive system and so acted as a natural contraceptive had to be rediscovered in the twentieth century and rescued from the dustbin of old wives' tales. How had something considered so obvious in Leonardo's time become by our own time so discredited? Part of the answer may lie in the social transformations of the industrial revolution in the West. The demise of wet-nursing as a common practice, the annexation of obstetrics and postnatal care from the domain of midwifery by male physicians, the spread of pasteurization and refrigeration and the ready availability of cow's milk and other breastmilk substitutes, and even (as we shall see) the improved nutritional status of the working classes and rural peasantry all helped to remove long periods of amenorrhea during lactation from women's common experience. But an intellectual shift occurred as well in the way social and medical scientists thought about human fertility. A new paradigm emerged that precluded any significant physiological regulation of normal fertility, traceable ultimately to the ideas of Thomas Malthus.

Malthus was an English parson of the late eighteenth and early nineteenth century credited as an important contributor to many areas of modern thought, including economics, evolutionary biology, ecology, and of course population dynamics. His famous treatises on "the principle of population" were composed in reaction to a dominant atmosphere of progressive, scientific optimism, if not to say arrogance, about the human condition prevalent at the crest of the Enlightenment. Leading social philosophers such as William Godwin in England and the Marquis de Condorcet in France had proclaimed that scientific progress would lead inevitably to a utopian future in which wisdom and justice would reign, the human life-span would extend nearly indefinitely, and birth rates would fall to the barest minimum. Human beings would approach the status of passionless, immortal angels, uninterested in procreation. To Malthus these ideas seemed absurdly uninformed about both human nature and the quasi-mathematical realities of population biology. He encapsulated these realities in what he called "the principle of population": that human populations, like those of other animals, have an innate tendency to grow exponentially unless held in check by specific restraints. The number of births in a population in a given time, he argued, is proportional to the size of the population. The more people you have, the more births you get. The absolute number of people in a population increases at an accelerating rate, like money in the bank at fixed interest, not at a constant rate, like the accumulation of a child's allowance in a piggy bank. Malthus was also mathematician enough to know that exponential growth of this kind would eventually exceed any limit one might care to fix.

The principle of population is a bit like another product of the Enlightenment, Newton's laws, the first of which stated that a body will continue its state of motion or rest until acted upon by another force. The significance of this universal law lies in the fact that the perpetual straight-line motion it postulates never occurs. It is the deviation from the postulated motion that allows the action of forces like gravity to be inferred. Similarly with the principle of population: it specifies the way populations *would* grow if nothing happened to prevent them from expressing this basic tendency. Because it is impossible for exponential growth to continue indefinitely, the action of such checks must be unavoidable. Something must eventually stop population growth. "When" and "what" may be debatable, but not, Malthus insisted, "if."

Malthus reasoned that there are only two basic ways in which population growth can be halted: either death rates increase or birth rates decrease, or some combination of the two. Emigration might help a local population, but not the global population. There is no other escape. So far so good. But at this point Malthus introduced into his logic a dichotomy that has persisted in the thinking of social scientists for more than two centuries. Natural checks on population growth would occur through increases in mortality: plague, pestilence, and famine, augmented perhaps by war. These are the causes of increased mortality that would eventually bring a halt to human population growth as the carrying capacity of the land was exceeded. However high one imagined the upper limit on the availability of life's necessities—food, water, room to stand, air to breathe—exponential growth would eventually surpass them all and increased mortality would be the necessary result. Not the utopia Godwin and Condorcet had imagined.

The other possibility, reduced birth rates, Malthus argued, would not occur naturally but would have to be the result of "moral restraint," a conscious check on sexual activity exercised by individuals or societies (through the regulation of marriage, for instance), or both. Fertility in

Malthus's thinking was not subject to natural control, like mortality. Indeed, if it were, if birth rates naturally declined as populations grew and resources became more limiting, then exponential growth would also slow and the principle of population would not hold. Such a natural decline in birth rates was in keeping with the vision of Godwin and Condorcet and exactly what Malthus was arguing against. Instead of allowing for such natural variation in fertility, Malthus argued that the biological potential for reproduction in humans, as in other animals, was high and invariant. If our reproductive biology were unchecked by individual and societal restraint, we would breed at a prolific—and ultimately horrific—rate.

Malthus underscored the distinction he was making by creating different names for the two possible checks on population growth: "positive checks" referred to those increases in mortality that nature would provide if the "preventive checks" of moral restraint were not exercised by individuals and societies. We could not expect to have both long life and low fertility without conscious effort, whatever Godwin and Condorcet might imagine. Although often contested and controversial, the Malthusian argument has persisted as an important strand in our modern intellectual fabric. Just as persistent, although less often acknowledged, has been the Malthusian dichotomy between "natural" mortality and "social" fertility. This dichotomy implied that there were no mechanisms for "natural" fertility regulation through biological responses, pathologies excepted. A contraceptive effect of breastfeeding would represent such a mechanism of natural fertility regulation. Because it did not fit the Malthusian paradigm, it seems to have been forgotten or ignored.

In the mid-twentieth century concern about population growth and the specter of global overpopulation kindled a new and widespread desire among social scientists to understand fertility regulation. Several important efforts were made to understand the causes of fertility differences between human populations as policy planners sought to develop strategies to bring global population growth under control. Lists of potential contributing factors were drawn up, but lactation was never included. In an influential UNESCO volume, *Culture and Human Fertility*, published in 1954, cultural prohibitions on intercourse during lactation (a form of social behavior that fits comfortably within Malthus's category of "moral restraint") are presented as potentially important regulators of fertility while a direct, physiological, contraceptive effect of lactation is not mentioned.

Particularly striking is the absence of any role for a suppressive effect of lactation on female fecundity in Kingsley Davis and Judith Blake's landmark paper, "Social Structure and Fertility: An Analytical Framework," published in 1956. Davis and Blake came as close as any social scientists of the day to allowing biology into the analysis of fertility regulation. They used what they called "intermediate variables" to illuminate how social factors affect fertility. (The assumption that it *is* social factors that are responsible for fertility variation is itself a product of the Malthusian legacy, of course.) Any social variable that is to influence human fertility, they argued, must do so by affecting one or more of a finite set of "intermediate variables." The intermediate variables fell into three major categories, reflecting the essential biological realities of procreation. A social variable must affect the formation or dissolution of a reproductive union between a man and a woman, the probability of conception given such a union, or the probability of a successful birth given conception. Davis and Blake then broke each of these three major categories into several individual intermediate variables. Conspicuously missing, from the perspective of the present discussion, is any mention of lactation, despite the fact that the essentially biological nature of reproduction was used as an organizing principle. The closest possible category listed is "fecundity or infecundity from involuntary causes." But the same category is later referred to as "sterility" and the discussion in the text makes it clear that Davis and Blake had permanent, pathological infecundity in mind, such as might result from venereal disease.

It is perhaps not surprising, given their unconscious adherence to the Malthusian paradigm, that the conclusion Davis and Blake draw from applying their analytical framework to an analysis of global fertility variation is also classically Malthusian. Two major factors are responsible for the bulk of fertility variation, they conclude: marriage patterns and contraceptive use. Malthus's religious precepts prevented him from viewing contraception favorably. But if we allow contraceptive use to represent a modern version of "moral restraint" at the individual level while marriage patterns represent the societal level, Davis and Blake's view of fertility regulation is exactly the same as Malthus's.

The use of "intermediate variables" in fertility research has remained centrally important in the field of demography, although currently the most frequently used term for these variables is "proximate determinants." The demographer John Bongaarts is usually credited with revitalizing the use of proximate determinants in analyzing fertility variation, and others have followed his lead. As an analytical tool, however, the use of such determinants has a serious bias. The bias occurs because there is more than one possible list of proximate determinants. As noted above, lactation was not even included as an intermediate variable in Davis and Blake's list. Of course, a variable that is not on the list, or that is buried with many other things in a catch-all category like "involuntary infecundity," is unlikely to emerge in an analysis as individually important. Bongaarts, working thirty years after Davis and Blake, appreciated the contraceptive effect of lactation and included it on his list of proximate determinants. In his analysis lactation emerges as a factor of major importance in determining population fertility levels, nearly equivalent to marriage and contraceptive use. On the other hand, Bongaarts's analysis does not include variation in the frequency of ovulation as a proximate determinant and so does not identify it as an important factor in fertility differences. Kenneth Campbell and James Wood, working later than Bongaarts, do include ovulatory frequency as a proximate determinant on their own list. And presto! It emerges as a runner-up to lactation.

As the list of proximate determinants is changed to include more physiological factors, those factors are discovered to be more important. In Davis and Blake's original list of thirteen intermediate variables, only two can really be classified as physiological while the other eleven are behavioral or social. Fully five of the thirteen variables refer to some aspect of marriage patterns. By the time we get to Campbell and Wood's list of proximate determinants, nine of thirteen can be classified as physiological and only one reflects marriage patterns. With the proximate determinants approach, what you see is largely what you get. As an analytical method it brings different pathways of fertility variation into and out of focus, but it is a lens that is never free of distortion.

GOING NATURAL

The legacy of Malthus, alive and well in the twentieth century, effectively obscured the role of lactation in regulating human fertility through the implicit assumption that physiological variation in human fecundity was unimportant. This assumption remained unchallenged in part because it was so difficult to test. How could one strip away the powerful effects of societal influences and individual choices to examine the variation that might remain at the physiological level? In 1961 the French demographer Louis Henry made a dramatic breakthrough in tackling this difficult problem, a breakthrough that not only exposed lactation as a major regulator of human fertility but also exposed the entire landscape of physiological variation in human fecundity to serious investigation.

Henry's breakthrough was based on defining a new concept, "natural human fertility." Henry defined natural fertility as fertility that is not consciously limited on the basis of the number of children already born. The pace of childbearing under natural fertility might be faster or slower, but it would show no evidence of changing dramatically when a certain target family size was achieved. The antithesis of natural fertility, as Henry defined it, was therefore not "unnatural fertility," but "controlled" fertility. This definition might strike many population scientists today as awkward, and indeed it has generated controversy over the years. Some might prefer to distinguish populations that use artificial birth control from those that do not rather than trying to determine the role of "deliberate choice" in limiting family size. But Henry cleverly constructed his definition to be used empirically. Its implementation depends on an analysis of parity progression ratios, or the proportions of women who have had N births who go on to have N + 1. In a natural fertility population, this proportion declines only gradually as N increases, since by definition there is no change in reproductive behavior that depends on the number of children already born. In controlled fertility populations on the other hand, parity progression ratios drop rather abruptly as the mean desired family size is reached and surpassed. Using parity progression ratios, then, Henry could sort populations empirically into the categories of natural or controlled fertility. He managed to identify thirteen populations with reliable data that showed clear evidence of natural fertility. Today, thanks to the research inspired by Henry's approach, the list of natural fertility populations is much longer.

In his analysis of the available data on natural human fertility, Henry introduced a further constraint. He considered only the fertility of married women. Not only was this empirically convenient (since data on marital fertility are generally much more complete and reliable than data on extramarital fertility), but it eliminated the second major avenue of societal regulation of human fertility. Human fertility uninfluenced by marriage or conscious limitation was, in fact, human fertility without Malthus's preventive checks, without the influence of Davis and Blake's two primary intermediate variables. Such fertility, if Malthus and his intellectual inheritors were correct, should be uniformly high. What Henry found was quite different. There was a twofold range of variation in overall fertility levels between the natural fertility populations, a range that has expanded to more than fourfold in the decades since as the list of natural fertility populations has grown. The lowest natural fertility populations fall comfortably within the range of controlled fertility populations. The basic Malthusian assumption that human fertility would be uniformly high in the absence of deliberate societal and individual control was clearly incorrect.

Having exposed the degree of variation in natural human fertility, Henry proceeded to consider its source. He quickly ruled out the ages at which women began and ended childbearing as significant contributing factors. Differences in overall marital fertility in natural fertility populations must, therefore, be differences in the pace of childbearing. To examine this aspect of human reproduction Henry introduced a new analytical framework, one that emphasized the temporal nature of the reproductive process and time as a limiting resource. He argued that the interval between two consecutive births—the "interbirth interval"—could be decomposed into three components: the time from the first birth until the mother was "susceptible" to another conception; the interval from the resumption of "susceptibility" to the next conception; and the period of gestation. In addition to these three component subintervals Henry added the possibility of a miscarriage or fetal loss, which would extend an interbirth interval by causing some of the component subintervals to be repeated. Variation in natural human fertility, Henry reasoned, must be traceable to variation in one or more of these birth interval components. On the basis of evidence available at the time, he quickly eliminated variation in gestation length or the rate of pregnancy loss as likely candidates. Eliminating variation in the second component—the "waiting time" between resumption of susceptibility and the next conception—was more difficult. The resumption of susceptibility, after all, is not easily observed and certainly is not reported in the vital statistics of populations. Henry argued, however, that there was little evidence in his populations for variation in the interval from marriage to first conception, which was an analogous sort of "waiting time."

Having eliminated the other possibilities, Henry concluded that variation in natural human fertility must essentially be variation in the duration of the "non-susceptible" period following a birth. Resumption of susceptibility to conception was potentially the product of two different factors. Either it resulted from a resumption of sexual relations, or it resulted from a resumption of biological fecundity (which Henry equated with the resumption of ovulation). Although anthropologists had often cited evidence of postpartum taboos on sex, especially in sub-Saharan Africa, Henry was not persuaded. Such beliefs did not go far in explaining patterns of fertility in Africa, much less the rest of the world.

Instead, Henry argued that the practice of breastfeeding, by influencing the resumption of ovulation, was the principal regulator of human fertility.

This was a bold claim. The notion that lactation suppressed ovulation was still viewed with deep skepticism by many physicians and physiologists. Some nursing women went months without resuming menstruation, it was true, but others resumed with little if any noticeable delay. Even Henry acknowledged this difficulty: "Ovulation is resumed very rapidly among some women even though they are suckling their babies. With others some time elapses after confinement before it is resumed but still before the infant is weaned; and with still others, it is resumed only after weaning." Henry viewed this variability not as a refutation of the effect of lactation on natural fertility, but instead as the heart of the relationship. If we could understand why nursing women fell into different categories in terms of the speed with which they resumed ovulation, he reasoned, we would be well on our way to a full understanding of variation in natural human fertility. He also noted that such an understanding would, in its essence, be "physiological" rather than sociological, another challenge to the basic Malthusian paradigm of human fertility: "To accept the second explanation [the effect of lactation on the resumption of ovulation] is to admit that this physiological characteristic can be included among those which, for reasons of heredity or environment, vary from one population to another. Research in this field would help us to understand why there exists among populations such variability in natural fertility."

GETTING INTENSE

The fact that women usually do not resume menstruation immediately after giving birth was well known to clinicians and physiologists in Henry's day. However, it was not clear to what extent this temporary absence of menstrual bleeding was a residual effect of pregnancy or a current effect of lactation. Large studies conducted in the United States in the 1930s had found a mean duration of such postpartum amenorrhea among nursing mothers of only three to four months, scarcely longer than the same period of amenorrhea observed among women who did not breastfeed. In the late 1930s and 1940s, researchers turned to histological examination of endometrial tissue obtained from the uterus by biopsy to confirm resumption of ovulation. But as with the resumption of menstruation, the variation in timing between different lactating women was considerable. Isadore Udesky, one of the principal researchers in this field, when pressed by her colleagues, speculated that the "intensity" of lactation might somehow vary between women and that such variation could account for different delays in the resumption of ovulation. Perhaps it was not simply whether a woman nursed her baby or not, but how *intensely* she nursed it, that mattered.

During the postwar years information on the fertility of non-Western populations began to accumulate rapidly, and with it important support for the role of lactation in delaying the resumption of menstruation and ovulation. In a number of populations in Africa and Asia median durations of postpartum amenorrhea of a year or longer were reported. Women in these populations who did not nurse their infants at all, usually because the infant died soon after birth, resumed menstrual cycling much more rapidly than those who did breastfeed. But the variation between women who did nurse was still dramatic. Some resumed menstruation as quickly as the nonnursing women; others remained amenorrheic for two years or more. Could the "intensity" of nursing play a role in this variation as well?

In order even to address this question, the notion of nursing "intensity" had to be translated into something concrete and measurable. But what did it consist of? The mechanical force of the baby's sucking? The rate of milk production and consumption? The degree to which the infant's diet was supplemented by foods other than its mother's milk? There was, in fact, some evidence that supplementation mattered to the resumption of menstruation. Mothers in the United Kingdom who practiced "full" or "unsupplemented" breastfeeding remained amenorrheic longer, on average, that those who supplemented their babies with other foods. But considerable variation remained in menstrual resumption among "full" (as opposed to "partial") breastfeeding mothers. The "intensity" of nursing, if it was the key, had to be something that varied among fully breastfeeding women as well.

In the 1970s a solution to this puzzle was proposed that at once seemed physiologically sophisticated and intuitively appealing. The "intensity" of breastfeeding, it was proposed, was not a matter of the amount of milk produced or the extent of supplementation, but rather lay in the temporal pattern of breastfeeding. The clue had come from studies of prolactin release and its relationship to nursing. The role of prolactin in milk production was well known, but it was also suspected that high levels of prolactin suppressed the female reproductive system. Women with prolactin-secreting tumors were nearly always amenorrheic, and if their tumors were excised or brought under pharmacological control, their menstrual cycles resumed. A similar mechanism was suspected to underlie the amenorrhea of nursing mothers. Soon after accurate techniques for measuring prolactin were developed, J. E. Tyson monitored the levels of British mothers as they breastfed their babies in order to understand better the dynamics of prolactin release. He observed an acute response of prolactin to each nursing bout. When the baby began to feed, the mother's prolactin levels rose rapidly to several times their prefeeding levels. When the baby stopped nursing, the levels declined more gradually as the prolactin was cleared from the mother's bloodstream.

This pattern of release immediately suggested a relationship between the frequency of nursing episodes and the average levels of prolactin in a woman's bloodstream. If a mother nursed her infant at short intervals, her prolactin level might never decline all the way to the unstimulated baseline. Rather it would fluctuate about a high average level for long periods of time, like a balloon buoyed by brief but frequent puffs of air. In a mother who nursed at wide intervals, however, the prolactin level would have time to return to the baseline between nursing bouts. Her average level would thus be much lower, punctuated by brief surges at feeding times. Two women might, therefore, feed their babies the same amount of milk per day and nurse them for the same total amount of time. But if they differed in the *frequency* with which they nursed their infants their prolactin profiles might be very distinct. If prolactin did indeed act to suppress the female reproductive system, these women might differ significantly in the degree of suppression they experienced.

This interpretation of nursing intensity was based on a sophisticated physiological argument, but it also had strong intuitive appeal. The temporal pattern of nursing was, after all, one of the most obvious ways in which nursing mothers in the United States and the United Kingdom differed from their peers in the developing world in the 1950s and 1960s. Conventional wisdom and professional advice concurred in urging Western mothers to train their babies to accept broadly spaced feedings and to sleep through the night as soon as possible. Cultural sensitivities also discouraged public nursing. The plentiful availability of breastmilk substitutes and supplements made it relatively easy for Western mothers to go for long periods of time without nursing their babies. For women in much of the world, however, nursing occurred much more frequently, endorsed as proper maternal behavior rather than stigmatized. It is worth noting, however, that the ability of different frequencies of nursing to generate differences in average prolactin levels was *inferred* from the pattern of prolactin release described by Tyson, and not actually observed.

No sooner had the "nursing frequency hypothesis" been articulated than data to support it began to accumulate. A team of Belgian researchers in Zaire, for example, obtained random blood samples from ninety-seven lactating mothers together with information on the ages of their children and the number of times per day they nursed. When these data were compared to levels of prolactin measured in the mothers' blood, an association with nursing frequency was apparent. By the time the mothers were six months postpartum, the prolactin levels in mothers who nursed their babies four times a day or less were indistinguishable from the levels in nonlactating women. Women who nursed their babies six time a day or more, on the other hand, showed scarcely any decline in prolactin levels even by the time they were twelve months postpartum and were much more likely than mothers who nursed less frequently to still be amenorrheic on their babies' first birthdays. When the researchers turned their attention to mothers of older children, however, the correlation between nursing frequency and prolactin disappeared. Few mothers reported nursing their children as often as six times a day in their second year, and no relationship could be observed between nursing frequency and either prolactin levels or menstrual status. It remained unclear whether frequent nursing was capable of sustaining prolactin levels into the second postpartum year. One year's delay in the resumption of fecundity was certainly a substantial effect, but not enough to account for the range of variation in natural fertility that Henry had documented. Was nursing frequency indeed the answer to Henry's problem of the variable effect of lactation on the resumption of postpartum fecundity, or only part of the answer?

In 1980 the anthropologists Melvin Konner and Carol Worthman seemed to provide the missing piece of the puzzle by reporting on the nursing patterns observed among !Kung bushmen of the Kalahari Desert in Botswana. The !Kung were at the time one of the few remaining examples of a people subsisting entirely by foraging wild resources. Such a hunting-and-gathering lifestyle is assumed to have been typical of human beings for the majority of our evolution until the relatively recent development of agriculture and animal domestication. For this reason the !Kung have played a large part in theorizing about human evolution. They also represent one extreme of the spectrum of human nursing patterns. Konner actually observed the nursing behavior of twenty-two pairs of !Kung mothers and infants, timing nursing events with a stopwatch rather than relying on self-reports. The pattern that emerged was quite astonishing. The !Kung nursed their infants nearly four times an hour, throughout the day. As the women pursued their subsistence activities, they carried infants and young toddlers in a sling that allowed them free access to their mother's breast. As a result even children eighteen to twenty-four months old appeared to nurse "on demand." Such a pattern, Konner and Worthman reasoned, might well maintain prolactin levels at a chronically high level and keep ovarian activity suppressed into the third year postpartum. Cross-sectional analyses of ovarian steroids in a subsample of !Kung mothers seemed to support the notion. Levels of the ovarian steroids estradiol and progesterone in randomly drawn blood samples from sixteen nursing mothers rose with the age of the child, but in a pattern that was significantly correlated with the average length of the interval separating suckling bouts.

As compelling as they are, the studies in Zaire and Botswana have limitations. The data they provide are in many ways incomplete and ambiguous. Both sets of data are cross-sectional, rather than longitudinal. Despite the fact that the Belgian team connected the points on their graph representing prolactin levels at different infant ages for mothers exhibiting a given nursing frequency, there was no way to know whether individual women actually followed these trajectories. Perhaps the women who were nursing their infants four to six times a day at one year had nursed at even higher frequencies when their infants were younger. The causal arrow linking nursing frequency and prolactin might even point in the reverse direction from the authors' interpretation. Perhaps women with high prolactin levels produce more milk and are more likely than others to nurse their infants frequently. Perhaps women vary in their physiological capacity to sustain prolactin and milk production with the passing of time. Those with greater capacity might reasonably be expected to nurse older infants more often than those with less. Konner and Worthman in fact made no measurements of prolactin at all. They merely argued the consistency of their data with the supposition that high nursing frequency would lead to high prolactin levels even when the mothers were two to three years postpartum—this despite the fact that the Belgians had been unable to find such a relationship in Zaire. Konner and Worthman did find that ovarian steroid levels were correlated with the interval between nursing bouts for different !Kung women, but they were also correlated with the infant's age. The fact that a mother's steroid levels rise as her infant ages might be due to a declining frequency of nursing, or it might simply be due to the passage of time. Or again, the causal arrow could be reversed. If ovarian steroids undermine milk production, perhaps women who resume ovarian activity sooner produce less milk than others and nurse less frequently as a consequence.

THE LONG ROAD BACK

Many of the questions raised by the Zaire and Botswana studies can be addressed only with longitudinal data on individual women covering the period from parturition to the resumption of full fecundity. As daunting as the task of collecting such data might seem, several research groups have risen to the challenge. Alan McNeilly, P. W. Howie, and their colleagues at the University of Edinburgh conducted the most comprehensive and influential study in the late 1970s. The Edinburgh study involved thirty-seven mothers, twenty-seven of whom breastfed and ten of whom bottle-fed their babies from birth. These mother-infant pairs were followed longitudinally from the birth of the baby through the resumption of ovarian activity and ovulation in the mother. Mothers kept daily records of the duration, timing, and number of nursing episodes, the number of supplementary feeds (either formula or solid food), and periods of menstrual bleeding. Twenty-four-hour urine samples were collected each week for estimation of ovarian steroid production. Every two weeks a researcher visited the mothers to collect the accumulated information and samples and to draw a blood sample for prolactin determination. From these data an extremely detailed picture was compiled of the changes through time in mother-infant nursing behavior and its relationship to the hormonal profile of the mother. No other study has been as complete in the array of information collected.

The data from the Edinburgh study provide a picture of the unfolding interactions of nursing behavior and maternal ovarian function through time. Lactating women resumed ovarian function later than bottle-feeders, as is indicated by the average dates of first menstruation (32.5 versus 8.1 weeks postpartum) and first ovulation (36.4 versus 10.8 weeks postpartum). In breastfeeding mothers the frequency of ovulation increased with time and as lactation was phased out, occurring in 45 percent of the first menstrual cycles during lactation, 66 percent of subsequent cycles during lactation, 70 percent of the first cycles after weaning, and 84 percent of subsequent cycles after weaning. This gradual resumption of full ovarian activity contrasted sharply with the rapid resumption of activity in bottle-feeders, among whom 94 percent of second and subsequent cycles were ovulatory. Even when ovulation resumed among lactating women, many of the resulting ovarian cycles were characterized as "luteally deficient," with low or abbreviated progesterone rises in the luteal phase of the cycle after ovulation.

The results also supported the association of nursing patterns, prolactin levels, and ovarian function. Thirteen of the twenty-seven breastfeeding mothers had their first ovulation while still lactating; the other fourteen had suppressed ovarian function throughout the period of lactation. Those who resumed ovulating during lactation had all recently changed their pattern of breastfeeding, reducing nursing frequency to less than six bouts a day and total nursing time to less than sixty minutes a day, and had introduced at least two supplementary feeds per day. In association with this change in feeding patterns, average prolactin levels at first ovulation had fallen well into the range shown by nonpregnant, nonlactating women.

When the subjects were divided into three roughly equal groups—those who resumed ovulating before thirty weeks postpartum (early), between thirty and forty weeks postpartum (middle), or later than forty weeks postpartum (late)—different trajectories of prolactin and nursing behavior were apparent as well. The early ovulation group nursed with equal frequency but for significantly less total time than the other two groups, introduced supplementary foods more rapidly, and abandoned night-time nursing sooner. The early group also had prolactin levels that were consistently lower than those of the other groups, with average values reaching the nonpregnant, nonlactating level by twenty weeks postpartum. Members of the late ovulating group, in contrast, nursed more in total time at twenty weeks postpartum than the early ovulating group had done at four weeks, maintained a high nursing frequency (more than four times a day on average) for over thirty weeks, and introduced supplementary foods much more slowly. In addition, 85 percent of the late ovulating group continued the regime of night-time nursing through forty weeks postpartum, after all the infants of the early ovulating mothers had been fully weaned. Prolactin levels in the late ovulating group were nearly twice as high as those of the early ovulating group at four weeks postpartum and did not fall into the nonpregnant, nonlactating range until between thirty and forty weeks postpartum. The middle ovulating group fell nicely in between the early and late groups on all measures.

The richness of the Edinburgh data was most apparent in the longitudinal profiles that were constructed for individual mother-infant pairs. These profiles depicted the simultaneous trajectories of nursing patterns, prolactin levels, steroid levels, and menstruation. These plots drew attention to the importance of the introduction of supplementary food into the infant's diet. Howie and McNeilly argued that it is this event that changes the pattern of nursing in Scottish women, setting in train the physiological changes leading to the resumption of fecundity.

The Edinburgh data seemed finally to provide a firm basis for what instantly became the leading hypothesis regarding postpartum amenorrhea. Henry had asked why some lactating women resumed menstruating early and others late. Udesky had speculated that some aspect of the "intensity" of nursing might explain the difference. At first no one had been sure what that "intensity" consisted of or how to quantify it. But now there was an answer that was both theoretically satisfying and methodologically tractable. The "intensity" of nursing was reflected in the *frequency* of suckling. It could be measured with a stopwatch. The frequency of suckling in turn was linked to the average level of prolactin in the mother's blood, and the prolactin level was directly responsible for suppressing ovarian function. Everything seemed to make sense and to account for the available data. Western women typically had short periods of amenorrhea during lactation because they nursed their infants so infrequently. If they nursed as frequently as a !Kung mother (it was assumed) they too could remain amenorrheic for years! Variation in the length of postpartum amenorrhea within a population was probably also a result of variation in nursing frequency, as was the variation between populations. After all, it had been shown that the median duration of postpartum amenorrhea across populations was correlated with the median duration of breastfeeding. Presumably populations with early weaning were also populations in which nursing frequency declines early.

Howie and McNeilly's colleague at Edinburgh, Roger Short, cast the entire theory into an evolutionary framework that made it even more compelling. In the formative human past, he argued, nursing patterns like those of the !Kung would have been the norm. A baby would have had unrestricted access to its mother's breast, nursing on demand through the day and night. When the child's demand for maternal milk declined, its frequency of nursing would decline. This change would send a signal to the mother's reproductive system that it was safe to resume normal fecundity without the danger of having to "metabolize for three." This feedback between infant nursing frequency and maternal fecundity, Short argued, had evolved because it optimized birth intervals naturally. It prevented births from occurring at dangerously close intervals, while allowing the birth interval to adjust to environmental circumstances. If the environment did not offer sufficient alternatives to breastmilk for a young child to grow on, the demand for maternal milk would persist for a longer time postpartum and the birth interval would be consequently longer. If, on the other hand, alternative foods allowed an earlier transition from breastmilk, nursing frequency would decline more rapidly with time and birth intervals would be shorter.

So satisfying was the "nursing frequency hypothesis" that it was rapidly embraced by biologists and social scientists alike. In 1988 an international conference was convened in Bellagio, Italy, under the joint auspices of the Rockefeller Foundation, the World Health Organization, and Family Health International. The express purpose was to reach "a consensus about the conditions under which breast-feeding can be used as a safe and effective method of family planning." In less than three decades since the publication of Henry's paper the contraceptive effect of lactation had emerged from the closet of old wives' tales to become a principle of population policy. The hypothesis was so enthusiastically embraced, in fact, that it has been difficult for proponents to recognize contradictory data, much less to conceive of alternative hypotheses.

BEARING THE LOAD

Nevertheless, a strong alternative hypothesis explaining the contraceptive effect of lactation can be advanced, and the available data appear to support it over the nursing frequency hypothesis. The alternative still supposes that lactation suppresses female fecundity in a way that correlates with the "intensity" of lactation. Instead of nursing frequency, however, the "relative metabolic load" of lactation is viewed as the best measure of nursing intensity. Relative metabolic load represents the proportion of the mother's metabolic budget that is devoted to milk production. It is important to realize that this varies both with the absolute cost of milk production and with the amount of metabolic energy available to the mother. In a mother with a constant metabolic budget, the relative load of lactation decreases as milk production decreases. Between women, however, or for the same woman at different times, a given level of milk production may represent a greater relative metabolic load when her overall energy budget is low, or a lesser relative metabolic load when her overall energy budget is high. Not only does this hypothesis fit the available data better than the nursing frequency hypothesis, it is sounder theoretically, as I will try to show.

Cracks in the edifice of the nursing frequency hypothesis began to appear even while the consensus behind it was growing. But as suggested by Thomas Kuhn's theory of scientific revolutions, a hypothesis that is extremely satisfying can show a remarkable resistance to empirical challenge. The mechanism underlying the nursing frequency hypothesis, for example, no longer holds. Researchers now appreciate that prolactin does not itself cause the suppression of ovarian function observed in nursing women, as was once thought. While recognizing this fact, however, adherents of the nursing frequency hypothesis have suggested that the frequency of nursing it is still a valuable index of the "intensity" of lactation. After all, it is still true that high prolactin levels are associated with amenorrhea. Prolactin may not cause the amenorrhea, but it appears to be a marker for a physiological state that *does* cause amenorrhea. Maybe so, but it no longer appears that prolactin levels reflect nursing frequency well either, as was originally suggested by Tyson's data. McNeilly's group recently conducted a new study of nursing mothers that involved continuous, twenty-four-hour video monitoring and blood sampling in a clinical research unit. No clear correlation could be observed between nursing bouts and prolactin elevations. A study in Nepal among rural Tamang mothers also failed to demonstrate a significant difference in prolactin levels before and after nursing bouts. In the Tamang case prolactin levels appear to be sustained at high levels without any acute response to nursing. In Edinburgh, especially among the mothers of older infants, levels appear to remain quite low without acute response to nursing. Prolactin levels may perhaps provide an index of the "intensity" of lactation, but not, apparently, a good index of nursing frequency.

Data from the field also often fail to support the nursing frequency hypothesis. For example, an extensive effort was undertaken in Bangladesh to collect data on nursing frequency and the resumption of menstruation from a large sample of women. The results failed to show any relationship between nursing frequency and the return of menstruation. So convinced were the authors of the study of the validity of the nursing frequency hypothesis, however, that they assumed their own data were somehow flawed or incomplete. Similarly, a careful study in Australia conducted by Roger Short and his colleagues failed to find any relationship between nursing frequency and the resumption of ovarian function. Even the original Edinburgh data do not support the hypothesis as strongly as is often assumed. In the individual longitudinal plots, for example, the introduction of supplementary foods often leads to a reduction in the total time spent nursing, but not in nursing frequency. In the new Edinburgh study that involved twenty-four-hour monitoring of nursing behavior and hormone levels, variation in nursing frequency did not account for variation in the resumption of ovarian function, but variation in the introduction of supplementary food did.

Comparisons between populations must be made carefully in order to discriminate between the nursing frequency hypothesis and the relative metabolic load hypothesis. Populations in which frequent and prolonged nursing is the norm are also often populations characterized by moderate undernutrition, making the relative metabolic load of lactation greater. Imagine a two-by-two grid with the columns labeled "high" and "low" nursing frequency, and the rows labeled "high" and "low" relative metabolic load of lactation. Most populations tend to fall either in the upper left quadrant (high nursing frequency and high relative load) or the lower right quadrant (low nursing frequency and low relative load). The fact that the populations in the upper left quadrant tend to have long durations of postpartum amenorrhea while those in the lower right have short durations does not help determine whether nursing frequency or relative metabolic load is responsible for the difference. Populations in the other two quadrants are harder to come by. Populations with low nursing frequency but high relative metabolic load of lactation may not exists at all (or not for very long when they do). But there are populations that demonstrate high nursing frequency and low relative metabolic load of lactation. These populations can potentially help disentangle the effects of these two variables.

Breastfeeding behavior, as an aspect of cultural behavior, often shows a considerable amount of inertia and resistance to change in traditional societies undergoing rapid modernization. As a result, very frequent and prolonged lactation is sometimes practiced in populations that are no longer as nutritionally stressed as they may have been in the recent past. In these cases, nursing frequency and relative metabolic load are no longer confounded as variables and useful comparisons with other populations can be made. For example, in a study of the Yolgnu, an Australian aboriginal population living on a government settlement with ample food, Janet Rich found that the well-nourished Yolgnu nurse their infants just as often as the !Kung but have median durations of amenorrhea half as long. The Amele in lowland Papua New Guinea nurse their infants with a high frequency as well, as high as in the !Kung or neighboring groups in highland Papua New Guinea. Their nutri-

A Toba woman with her seven children and two grandchildren (next to their mother in the front row).

tional status and body weight, however, is much higher. Carol Worthman and her colleagues found that their prolactin levels fell rapidly postpartum despite frequent nursing, and that their duration of amenorrhea was also brief compared to that in highland groups or the !Kung. The Toba are a native population living in northern Argentina. They formerly lived as hunter-gatherers much like the !Kung, but are currently living on government settlements. Like the Yolgnu, they nurse their children in traditional patterns with high frequency, averaging over three bouts an hour throughout the day. Their nutritional status, however, is very high, with body weights that are heavy for their heights by international standards. The average duration of amenorrhea among the Toba is only 10.5 months and the average interbirth interval only 24.8 months. Despite high frequency nursing, then, the Toba have children at dangerously close intervals.

Perhaps the most telling data come from the dietary supplementation studies in the Gambia referred to previously. Recall that nutritional supplementation of pregnant and lactating mothers proved surprisingly ineffective at raising birth weight or increasing milk production. Rather, the supplemental calories seemed to go to the mother, relieving her of the necessity of depressing her own metabolism to free up energy for her offspring. An additional, unexpected effect of the supplementation was also observed: a more rapid decline in prolactin levels postpartum, a more rapid resumption of menstruation, and a shortening of the interval to the next birth by nearly a third. When supplementation occurred during pregnancy as well as lactation, the decline of prolactin was even more rapid, and the intervals to the resumption of menstruation and the next conception became even shorter, than when supplementation occurred during lactation alone. In struggling to understand this result the researchers in the Gambia speculated that the improved nutritional status of the mothers might have lowered their nursing frequency. But there are no data to support that speculation. In fact, since milk production was monitored and found not to be affected in supplemented mothers, there is little reason to expect that nursing frequency did change, especially given its resistance to change in response to improvements in nutritional status in other populations. The alternative explanation is that the supplementation program effectively raised the overall metabolic budget of the mothers, thus lowering the relative metabolic load of lactation.

If the relative metabolic load hypothesis is correct, then prolactin may still be a good index of the "intensity" of lactation. It would fill this role not because of any supposed relationship to nursing frequency—a relationship that does not seem very tight in any case—but because of prolactin's direct role in "tipping" maternal metabolism toward milk production. The lower the overall metabolic budget, the higher prolactin levels have to be in order to sustain an adequate flow of energy to milk production. When metabolic energy is abundant, a much lower level of prolactin is sufficient. The Gambian data confirm this, since differing levels of prolactin support the same level of milk production under differing metabolic budgets.

The literature on lactation and reproduction is too large to sift through study by study. However, virtually all the data that can be cited to support the nursing frequency hypothesis also support the relative metabolic load hypothesis. This is true because the two principal variables are so often confounded. All other things being equal, a higher nursing frequency is usually an indicator of a higher relative metabolic load of lactation. This is true whether the contrast is between the mother of a younger infant and the mother of an older infant, between a "fully breastfeeding" mother and a "partially breastfeeding" mother, or between a high nursing frequency, low nutritional status population and a low nursing frequency, high nutritional status population. Where the data do not fit the nursing frequency hypothesis, as in the examples above, they can usually be explained by differences in metabolic budgets and hence relative metabolic load.

Theoretically, the relative metabolic load hypothesis is more satisfying as well. The evolutionary argument advanced by Roger Short assumes that nursing frequency will be a reliable signal of the energetic requirements of the current infant. This assumption in turn is based on the notion that nursing will be "on demand," that the baby is "in the driver's seat" in determining the frequency of suckling. Whether this has ever been true of any population, such as the !Kung, is debatable. It is increasingly clear that it is not true for most contemporary populations. Nursing patterns are highly structured by "opportunity" as well as by "demand." Competing demands on the mother's time, the availability of surrogate caretakers, as well as cultural and personal notions of what the "correct" pattern of breastfeeding is, have all been found to have a strong influence on the frequency of nursing where they have been studied. Even among nonhuman primates there are strong differences in maternal "temperament" that affect nursing patterns. Some mothers are very indulgent of their infants' demands to nurse, others are very resistant. It seems quite a leap of faith to assume that nursing frequency is a sufficiently reliable and robust signal of an infant's energetic requirements to serve as the basis for optimizing birth intervals. The total amount of milk consumed per day might be a better indicator, and one that would more closely reflect relative metabolic load.

But as noted above, relative metabolic load is not only a function of the amount of milk produced, but of the metabolic budget available to support it. Relative metabolic load depends on maternal condition in a way that nursing frequency does not. In most populations with long durations of lactation, menstruation usually resumes while milk production continues, albeit at a low level. Often, as in the Edinburgh studies, a shift in the infant's diet away from reliance on breastmilk toward a greater consumption of solid foods precedes the resumption of menstruation. But the point at which the cost of milk production has fallen sufficiently to free up energy for a new conception will depend on the mother's total metabolic budget. At a given low level of milk production, a mother with a high metabolic budget may have sufficient energy available to begin another pregnancy while a mother with a low metabolic budget may not. Even though their offspring may be nursing with the same frequency and receiving the same amount of milk per day, resumption of fecundity may make sense in the one case and not in the other. A physiological mechanism for suppressing fecundity based on relative metabolic load would make this distinction. A mechanism based on nursing frequency would not.

If the relative metabolic load hypothesis is superior, in terms of both its theoretical consistency and its empirical support, why did the nursing frequency hypothesis gain so much support in its stead, and why do so many researchers continue to cling to this hypothesis today? Why did researchers ever think that the "intensity" of lactation could be better measured with a stopwatch than with a calorimeter? The elegance of the prolactin connection as originally conceived is probably responsible for fixing attention initially on the temporal pattern of nursing as a key variable. The fact that such a mechanism made theoretical sense added to its appeal. Equally important, however, may have been the fact that the relevant variable was easily measurable in practical terms. No particularly invasive or disruptive procedures were required to gather information on nursing frequency. Recall and simple, timed observation provided the data for all of the important studies. By the time the direct causal connections between nursing frequency and prolactin levels, and between prolactin levels and ovarian suppression, began to erode, the notion that nursing frequency was an important measure of the "intensity" of lactation had already been established, and it persisted without the support of the mechanistic scaffolding that had been present when it was first erected.

An additional appeal of the nursing frequency hypothesis lies in its compatibility with the Mathusian paradigm that located fertility variation firmly in the social, rather than the biological, domain. Even if lactational suppression of female fecundity relied on physiological mechanisms, to the extent that nursing frequency served as the key variable, this source of fertility variation, too, could be considered ultimately subject to individual and cultural control. Whether consciously or not, individuals and cultures were "choosing" to follow patterns of breastfeeding that had certain consequences for fertility. If they made different choices, they could enjoy different consequences. In the strong form of the nursing frequency hypothesis, a Boston woman could "choose" to experience postpartum amenorrhea as long as a !Kung woman by "choosing" to nurse her infant at fifteen-minute intervals around the clock for two or three years. Similarly, a !Kung woman could "choose" to have a period of amenorrhea as short as that of a lactating Boston woman by nursing only every four hours and eliminating nighttime feedings as soon as possible.

Ever since Malthus there has been something very comforting to social scientists in the notion that human fertility is, pathology aside, simply a function of human behavior. Initially this position was supported by the assumption that there was no meaningful variation in physiological fecundity. In the face of Henry's refutation of this assumption, the nursing frequency hypothesis allowed for the next best thing: there may be meaningful variation in fecundity, particularly postpartum fecundity, but it is entirely behaviorally mediated. The relative metabolic load hypothesis challenges the notion that fecundity is ultimately so plastic and easily manipulated. Postpartum fecundity is not simply a matter of choosing when and how often to nurse an infant, but a matter of allotting available metabolic resources to competing objectives. Behavior certainly plays a role. Particularly when abundant alternative foods are available a mother may choose whether to nurse her baby at all and if so, how exclusively. She may choose how quickly to shift her offspring away from breastmilk to other foods. She may have some flexibility as well in determining her own nutritional intake and energy expenditure on other activities. But she cannot, simply by willing it, override her own body's decisions on metabolic energy allocation. Nor can she, by willing it, create metabolic energy that isn't available. In the end the tension between a woman's own metabolic requirements, those of her infant, and the potential requirements of another pregnancy allows for only so much give and take.

Unfortunately, the "social" and "biological" perspectives on variation in fecundity became dramatically polarized by a set of hypotheses proposed by Rose Frisch and their reception by the community of demographers and social scientists interested in population dynamics. We will examine Frisch's hypotheses is considerable detail in the next few chapters. But in brief, she introduced the idea that female fecundity, and to a large extent historical human fertility, is primarily determined by the amount of fat a woman carries on her body. Her views have been subject to serious challenge from many quarters, but a highly visible exchange of articles in the journal *Science* between Frisch and John Bongaarts served to polarize subsequent debate. On the one hand was Frisch, arguing that nutritional status, and especially fatness, was the key determinant of female fecundity with a powerful effect on the fertility patterns not only of primitive societies but of historical Europe as well. On the other hand was Bongaarts, arguing that the effect of nutrition was trivial compared to the effect of lactation in determining the fertility levels of those same populations. "Nutrition" was thus set in opposition to "lactation" as a competing explanation for variation in female fecundity. Once such a strong dichotomy has been established it tends to distort all subsequent distinctions in its own image. In the minds of some, a hypothesis like the relative metabolic load hypothesis becomes categorized as a version of the Frisch position while the nursing frequency hypothesis becomes categorized as a version of the Bongaarts position. Of course, this distortion is grossly unjust. Both the relative metabolic load hypothesis and the nursing frequency hypothesis assume that lactational suppression of female fecundity is the primary determinant of variation in natural fertility. They differ instead in the way in which they understand lactational suppression of female fecundity to be mediated.

The nursing frequency hypothesis and the associated Mathusian paradigm of socially determined fecundity are also viewed by some as more politically correct than any alternative that views female fecundity as physiologically determined. James Wood, for example, has suggested that demographers, by virtue of their deeper engagement with practical issues of population and public policy, are more inclined to see the merits of "lactation" as a determinant of natural fertility than are physiologists, whose distance from the same issues leads them to favor "nutrition." He notes darkly that demographers worry "privately" that the idea that maternal nutrition affects fertility carries disturbing policy implications, that adherents of this view would be opposed to improving the nutritional status of undernourished populations for fear of exacerbating population growth. Thus not only does he perpetuate the distorted opposition of "nutrition" and "lactation" as fertility determinants, he saddles the "physiological" view with the onus of an irresponsible social agenda.

The relative metabolic load hypothesis does not, of course, lead by necessity to the policy implications Wood suggests. Only to those who feel they have the right or responsibility to control the reproductive lives and population growth of others would such an implication occur. However, ignoring the importance of relative metabolic load in determining the contraceptive effect of lactation could itself lead to irresponsible policy conclusions. It would be wrong, for example, to suggest that traditional breastfeeding patterns will maintain the same contraceptive effect in a population that is undergoing a transition to higher caloric intakes, lower levels of physical activity, and lower disease burdens. Populations like the Toba, where birth intervals now average less than twenty-five months and completed family sizes approach eight offspring despite frequent nursing, show just how ineffective "nature's contraceptive" can become when ecological circumstances change. The policy implications of the relative metabolic load hypothesis are not that the nutritional status of populations like the Toba should be undermined. The implications are rather that their family-planning options should be broadened, and that the educational, economic, and health-care opportunities that empower individuals to make effective family-planning choices should be fostered.

The Bellagio Conference convened to determine "the conditions under which breast-feeding can be used as a safe and effective method of family planning" but adjourned with only a very weak recommendation: a woman can reasonably rely on lactation for contraception only as long as she is not menstruating. Even then there is a small but measurable chance of conception. Although it was generally agreed that more "intense" breastfeeding would generally extend the duration of amenorrhea, no particular pattern of nursing could be associated with a specific contraceptive effectiveness. In the end Henry's guess that the duration of amenorrhea during lactation depends on some physiological characteristic of the mother, a characteristic that varies with the environment, appears to be true. Nature's contraceptive helps to prevent a woman from being saddled with the untenable task of metabolizing for three by scaling its effectiveness to the mother's available metabolic energy. As her child grows and achieves a full metabolic independence, her own metabolic energy is released to potentially conceive again. The child, now physiologically on its own, must manage its own metabolic energy on the journey toward physical and reproductive maturity.

WHY GROW UP?

THE NEWLY WEANED CHILD faces two major physiological tasks: staying alive and growing up. In metabolic terms, these two tasks compete with each other to some degree. The energy devoted to skeletal growth, for example, is not available to use in mounting an immunological response to infection, and vice versa. We assume that natural selection has shaped human physiology to allocate energy and other metabolic resources in ways that enhance ultimate reproductive success and the contribution to future gene pools. Staying alive is clearly crucial to such an end, but why grow up? Why invest so much time and energy in the process of physical growth and maturation before getting down to the business of reproduction?

Of course many changes occur during this period besides physical growth and the maturation of physiological systems. Cognitive development, social learning, the accumulation of experience and skills, and the elaboration of social relationships beyond the nuclear family all are essential parts of growing up. It has been suggested by some that this process of learning and social development needs to be completed first in order for reproduction to have the best chance of success, and that the complex nature of this task in humans simply takes a long time. Although appealing on the surface, this argument is difficult to reconcile with data on ecological variation in the age of sexual maturity. In girls, for example, the average age of first menstrual bleeding (an indicator of cyclic ovarian activity) can vary from later than sixteen years to earlier than twelve years, or more than 25 percent, depending on nutrition, disease burden, and other ecological factors. If the effective constraint on reproductive maturation is the time necessary to garner sufficient wisdom, experience, skill, and social support, it should be less plastic and sensitive to other things. Instead it seems that the timing of reproductive maturation is linked to the process of physical growth and shares the same constraints. In environments where children grow up fast, in the physical sense, they mature early. This is true whether we are comparing historical periods within the same population or different populations. There are no counter-examples, no populations where children grow faster and mature later, or grow slower and mature earlier. Metabolic investment in growth *is* investment in reproductive maturation.

IN THE GROOVE

Childhood growth reveals clues to the way in which patterns of metabolic allocation have been shaped by evolution. For one thing, the process of physical growth appears to be very regular. By and large, children grow at a relatively steady rate from infancy to adolescence, so steady that their progress can be followed and projected on standardized charts. These charts represent the size distribution of children by age with lines drawn to connect given centiles of those distributions (such as the fifth, fiftiethth, and ninetieth) through the growth period. Individual children tend to grow in parallel to these centile lines. A child that is small for his or her age at five years tends to be at a similar place in the size distribution at ten years. Significant departure from the trajectory implied by these charts is usually a signal of some pathology (hormonal imbalance, serious illness, starvation, or anorexia for example), not an expression of normal variation.

In the past there was considerable debate over the appropriateness of using growth standards derived in one population (say, in Britain or the United States) to assess growth in another population (say, in Nigeria or Japan). After all, the distribution of sizes and the corresponding centiles can vary a great deal between different populations. The average ten-year-old in India may be much smaller than the average ten-year-old in France. Perhaps if we use the standards from one population to assess growth in another, we might inappropriately label normal growth as pathological. The consensus now is that the use of uniform standards can be justified (although population-specific standards can still be useful) and that the basic pattern of human growth is universal.

One reason for this conclusion is that growth standards developed for different populations all represent the same growth trajectories; it is only the relative centile labels that change. The fiftieth centile line in one population may correspond to the fifth centile line in another, but the lines do not diverge significantly. It is still expected that a child from either population that falls near a given centile line at one age will fall near the same line at a later age. A second reason for the validity of universal growth standards is the fact that population differences tend to disappear as differences in socioeconomic status and its ecological correlates (nutrition, health, and so on) do. Even in developing countries, where the average child may grow along the fifth centile of the standards from a developed country, children of the highest socioeconomic classes grow very close to the fiftieth centile, indistinguishably from their peers in the comparison population. Equally impressive is the evidence of the rapid rate at which immigrants from countries with "lower" growth trajectories to countries with "higher" growth trajectories converge on those higher trajectories. Second-generation immigrants can be very difficult to distinguish from the rest of a population on the basis of growth patterns. The differences in size and growth rate that remain for different human populations after ecological differences are removed are only a small fraction of the differences that are correlated with ecology. In the appropriate use of universal growth standards, however, we still have to distinguish between assessing community health and assessing individual health. Finding that the median growth trajectory for a given population falls low on a set of universal standards indicates that ecological circumstances are constraining growth in some way. Nevertheless we still expect healthy children in that population to grow in parallel to the standard growth centile lines. Divergence from those trajectories by any individual remains a reason to suspect pathology.

A second principle of human growth is that it is highly canalized, showing evidence of homeorhesis. Small day-to-day variations in ecological conditions do not produce corresponding variations in growth. The processes of cell division and cell growth that underlie physical growth tend to integrate metabolic investment over time, smoothing out shorter-term variation. Longer-term variations in metabolic investment can, however, have a significant effect on growth. Hence chronic differences in environmental conditions such as energy availability or disease burden can produce the sorts of differences in growth trajectories between populations discussed above. Longer episodes of variation in metabolic investment within an individual can also change a growth trajectory. These are the conditions that usually flag some pathology. The reduction in growth rate that accompanies weanling diarrhea is one example, as is the slowing of growth during prolonged episodes of infectious disease. For the same reasons chronic infections, such as many parasitic infections, may not be manifest as disturbances of normal growth patterns. The chronic presence of such infections may reduce the metabolic investment in growth and result in growth along a lower centile trajectory without producing an aberrant pattern.

A third principle of human growth is that skeletal growth is more tightly canalized than growth in weight. Short to medium-term variations in energy intake or energy expenditure, for example, will often produce changes in weight or rate of weight gain without affecting the rate of growth in height. In part this difference reflects the different processes underlying weight change and height change. Soft tissues can gain or lose mass by changes in cell size as well as cell number. Indeed, in many of the tissues that contribute substantially to short-term changes in weight, such as skeletal muscle and adipose tissue, cell number may be quite stable over time with reversible changes in cell size producing the variations in tissue mass. These tissues in fact serve as reservoirs for carbohydrates, lipids, and amino acids, helping to buffer the availability of important metabolic substrates from variation in intake and utilization. The design of these soft tissues includes this ability to cushion more essential physiological processes from the effects of environmental variation.

Growth in height is virtually entirely a function of skeletal growth, primarily of the long bones and secondarily of the vertebral column, pelvis, and skull. Skeletal growth is also driven by cellular proliferation and cell growth, but it includes a process of mineralization that renders it much less plastic and reversible than soft tissue growth. Bone can be resorbed; indeed, active resorption and redeposition of bone is a normal part of skeletal growth, the process by which the shape and density of bone is continuously adjusted. Bone also serves as a reservoir for the important minerals calcium and phosphorus. Resorption does not, however, usually reduce the length of the long bones. Compression of the vertebral column can cause a reduction in stature with age in older adults as bone density declines, but this is not a part of normal childhood growth. Ecological and metabolic disturbances, if they are severe or sustained enough, can slow bone growth or even bring it to a halt, but do not cause children to shrink. Overall, bone growth appears to be one of the processes that are buffered by the flexibility of soft tissue growth. In this sense growth in height appears to have a higher metabolic priority than growth in weight. Human physiology appears designed to sacrifice growth in weight to sustain growth in height, at least over the short to medium term. Conditions of chronic metabolic constraint on the other hand result in a slowing of both growth in height and growth in weight. For these reasons, an assessment of community health based on patterns of childhood growth usually discriminates low height for age from low weight for height. Low weight for height usually indicates a more recent or acute metabolic stress (such as undernutrition or disease). Low height for age indicates a more chronic stress, resulting in a lower growth trajectory.

The adjustment of growth trajectories to chronic ecological conditions is an example of developmental plasticity that is itself assumed to be adaptive. An individual growing up under conditions of chronically low energy availability (for example) may be better off growing slowly and being smaller as an adult. Slower growth will divert less energy from maintenance functions. Smaller adult size will also result in a lower average metabolic rate and lower maintenance costs. If these adjustments did not contribute on balance to evolutionary fitness, skeletal growth would be even more tightly canalized than it is and final adult size would not be affected by ecological conditions. Developmental plasticity represents a level of physiological response to ecological conditions that is much slower to take effect and much less reversible than physiological and behavioral adaptability. It is only appropriate to conditions that are very long term. But for the same reasons it is often overlooked by physiologists who study more rapid responses to shorter-term challenges. As we shall see, there are reasons to believe that human reproductive physiology also shows developmental plasticity as well as shorter-term and more reversible responsiveness to ecological conditions.

Of course not all variation in human growth is the result of ecological effects. Genetic factors play a significant role as well, as common sense and experience suggest. Many children are short because their parents are short, not because they are undernourished. Averaged across populations, these individual genetic differences become diluted but can still contribute to residual differences in average size after ecological effects are removed. There are many ways to assess the genetic contribution to traits such as growth and size, and they vary in sophistication and precision. One of the cruder but simpler methods involves the comparison of traits between monozygotic and dizygotic twins. Monozygotic ("identical") twins, resulting from the early cleavage of a single fertilized egg into two separate embryos, share all of their nuclear genes, whereas dizygotic ("fraternal") twins, resulting from two separate fertilized eggs, share on average only half of their nuclear genes. Both are raised together under nearly the same conditions. To the extent that monozygotic twins are more alike in a given trait than are dizygotic twins, the resemblance is assumed to be a result of their genetic similarity. A simple formula allows measurements of any trait on monozygotic and dizygotic twins to be converted into an index of this genetic effect, an index referred to as heritability. It is important to realize that heritability is not a "measure" of the genetic contribution to a trait; it is only an "index." It allows us to identify relatively greater, lesser, or similar levels of genetic contribution to various traits, but can tell us nothing more precise than that. The index of heritability varies between 0 and 1; the higher the index, the greater the genetic contribution.

The heritability of stature calculated in this way is quite high, on the order of 0.8–0.9. In contrast, the heritability of weight is much lower, around 0.4–0.6. Skeletal dimensions tend to have higher heritabilities, particularly those reflecting long bone lengths, such as the lengths of arms and legs. Head circumference also has a high heritability, around 0.8. A few size traits show different heritabilities in men and women. Shoulder breadth, for example, is somewhat more heritable in men (male monozygotic twins show greater convergence in this trait than female monozygotic twins). Pelvic breadth, on the other hand, is much more heritable in women. This may reflect the greater degree of selection pressure on pelvis size in women related to parturition. The age at menarche, or first menstrual bleeding, in girls is also very heritable, with monozygotic twins usually getting their first periods within a few months of each other. The heritability of menarcheal age is in fact as high as that of adult height, in the range of 0.8–0.9. Note that this does not mean that ecological factors do not influence menarcheal age. They manifestly do, just as they influence height. It does mean that there is a substantial genetic contribution to the determination of menarcheal age as well.

SPRINTING TO THE FINISH LINE

The period of growth in humans ends in a rapid burst known as the adolescent growth spurt. The steady and predictable course of childhood growth is suddenly accelerated to twice its previous rate or more over the course of a year or two, and then just as suddenly decelerates and comes to a halt. Sometimes the acceleration in growth is modest. Sometimes it is dramatic enough to be painfully obvious, at least to parents, as pant legs recede above the ankles and sleeves above the wrists of children who may be visibly taller when returning from summer camp than they were when they left home.

In addition to disrupting the steady pace of childhood growth, the adolescent growth spurt also disrupts many of the other regularities of growth. The timing of the spurt, for one thing, is quite variable, occurring years earlier in some children than in others. For this reason, the adolescent growth spurt is often easier to "see" in a plot of individual data than in a plot of average data for a group or population. Average data tend to smooth out individual spurts into a longer, slower pattern than that followed by any particular child. Ordinarily, taller children tend to have earlier spurts than shorter children do, so that the size differences between peers in a cohort tend to increase during the early part of the adolescent period and then to decrease later. Girls tend to experience the adolescent growth spurt earlier than boys do by an average of a couple of years, so that many girls are larger than their male peers for a while. Individual growth trajectories may shift during the adolescent growth spurt as well. A child who has been in the fiftieth percentile of height for age (for example) through childhood may shift to the seventy-fifth or the twenty-fifth centile during adolescence without implying any pathology. There is also evidence that adolescent growth patterns are more heritable than childhood growth patterns. Or to put it another way, childhood growth seems to be more strongly affected by ecological circumstances than adolescent growth, with genetic effects more strongly expressed in the latter. In communities where groups from different genetic backgrounds live under similar circumstances, the preadolescent children tend to be very similar in their sizes and growth patterns and to diverge, if at all, primarily during adolescence.

Physiologically, the adolescent growth spurt is a result of changes in the cellular processes that contribute to the growth of the long bones. In late childhood, a typical long bone consists of a shaft, or diaphysis, and secondary centers of ossification at the ends called epiphyses. Between the diaphysis and the epiphysis is an area of cartilage called the metaphysis, or growth plate. Cell division occurs in cells on the far side of the growth plate, the epiphyseal side nearer the joint surface of the bone. The newly formed cells then migrate through the cartilaginous matrix of the growth plate toward the shaft. As they make this journey they increase in size and pile up against the ossified bone of the shaft. Mineral deposition then occurs around the cells until they are replaced by solid bone. Lengthening of the bone thus occurs at its ends as the articular surfaces of the joints are pushed farther and farther from the center of the shaft. At adolescence it is this process that speeds up. The processes of cellular proliferation and growth speed up first, leading to an acceleration of long bone growth. But then the process of ossification accelerates to outstrip the process of cellular proliferation. The entire growth plate becomes mineralized, the epiphyses become fused to the shaft of the bone, and the opportunity for further growth is ended. The process of the adolescent growth spurt can be followed on x-rays of the articular ends of the long bones. There one can monitor the shrinking of the growth plates and the fusion of the epiphyses to the shafts. X-rays of the hand and wrist are particularly useful in this regard because they display many bones at once and can be obtained with minimal exposure of the subject to radiation.

Of course, the adolescent growth spurt is not the only physical event marking this period of life, or the most remarkable one. Accompanying the flourish with which physical growth ends are the development and maturation of adult reproductive characteristics. This process of reproductive maturation includes both the maturation of primary sexual characteristics (the functional activity of the gonads in producing gametes and hormones) and the maturation of secondary sexual characteristics (an array of physical features that are themselves products of the steroid hormones produced by the maturing gonads). Secondary sexual characteristics include enlarged genitalia, pubic and axillary hair, breast development in girls, beard growth in boys. Changes in body form and body composition, particularly increases in subcutaneous fat in girls and increases in muscle mass in boys, are also manifestations of reproductive maturation and the production of gonadal steroids. Under the influence of ovarian steroids, changes in the shape of the female pelvis also occur at this time, leading to a widening of the hips. Most of the secondary sexual characteristics of females are developed under the influence of estrogen hormones produced by the ovary, particularly estradiol, although the small amounts of androgen hormones (nineteen carbon steroids) produced by the ovary may contribute to the development of pubic and axillary hair. The secondary sexual characteristics of boys, on the other hand, develop primarily under the influence of androgen hormones, principally testosterone, produced by the testes.

Secondary sexual characteristics tend to develop in a reasonably orderly sequence. Breast development in girls usually begins before pubic hair appears. Menstruation usually begins after breast development has passed the preliminary stage and pubic and axillary hair has begun to develop. Increasing fat deposition usually occurs after menarche, and the broadening of the hips usually happens last of all, as breast development is approaching completion. In boys, enlargement of the penis and testicles usually begins first, followed soon by the appearance of pubic and axillary hair, a drop in vocal register, the appearance of facial hair, and increases in muscle mass. The sequence is not rigid in either sex, but development of characters markedly out of sequence can be a sign of pathology.

HURRY UP AND STOP

The coordination of primary and secondary sexual development is not surprising since the latter is in fact a result of the former. Individuals with specific rare pathologies of the endocrine system sometimes fail to undergo primary sexual maturation at all. In such cases secondary sexual maturation is absent as well. The coordination of reproductive maturation with the adolescent growth spurt is perhaps a bit more mysterious. The two processes are, in fact, usually quite closely synchronized. Menarche in girls, for example, usually occurs after the peak of the growth spurt as the rate of growth in height starts to drop toward zero. The anthropologist Franz Boas noted this relationship in the 1930s and suggested that "bone age" was a better predictor of reproductive maturation than chronological age. Standards of "bone age" can be developed from hand-wrist x-rays on large numbers of children. These standards are based on scoring the progress of epiphyseal fusion in the many bones visible in hand-wrist x-rays and arriving at a composite score to reflect the overall status of skeletal maturation. Different stages of secondary sexual maturation occur within much narrower ranges of bone age arrived at in this way than chronological age. Or put another way, grouping individuals on the basis of similar hard-wrist x-rays results in closer similarity of reproductive maturation than grouping on the basis of age.

Skeletal maturation is synchronized with reproductive maturation because it, too, is influenced by the production of gonadal steroids. The acceleration of growth in the adolescent growth spurt is largely a response to the production of androgens, while the deceleration and cessation of growth is largely a function of the production of estrogens. Androgens stimulate the processes of cellular proliferation and growth that are responsible for the elongation of the long bones, while estrogens stimulate the process of mineral deposition that results in ossification and eventual fusion of the epiphyses to the shaft of the bone. This effect of estrogen in stimulating mineral deposition is important in maintaining bone density in women. Low estrogen levels during lactation allow for mobilization of calcium and phosphorus in the production of milk. Low estrogen levels at other times, however, can result in a loss of bone density. After menopause this effect of low estrogen levels can contribute to the risk of osteoporosis. The relationship of estrogen production to ossification and fusion of the epiphyses helps to explain the normal occurrence of menarche during the final deceleration phase of the adolescent growth spurt. Menstrual bleeding is a sign that the ovaries are producing enough estrogen to promote endometrial growth in the uterus. The same estrogen that is responsible for causing menstruation to occur is also responsible for stimulating epiphyseal closure. Exogenous estrogen can be used to therapeutically bring growth to a halt in girls in whom ovarian maturation may be delayed for pathological reasons. Blocking the production of estrogen by the ovaries can allow growth to continue in girls with precocious puberty.

Even in boys, the accelerated ossification that brings skeletal growth to a halt appears to be an effect of estrogens. Although males produce far less estrogen that females do, they do produce enough to bring about this result. The importance of estrogen in ending the growth spurt in boys is compellingly illustrated by the case of a male born with a defective gene for estrogen receptors reported in *The New England Journal of Medicine*. Because his receptors did not bind the estrogen his body produced, his growth spurt continued for years and could not be artificially stopped by the usual estrogen treatment.

While estrogens are responsible for ending the growth spurt in males as well as females, androgens are responsible for initiating the growth spurt in females as well as males. Estrogens are produced in the ovary by a two-stage process. The first stage involves production of androgens, primarily testosterone and androstenedione, by cells outside the egg-containing follicles. These androgens are then transformed into estrogens, primarily estradiol, by cells inside the follicles in a process known as aromatization. Testosterone and androstenedione are thus precursors of estrogen, necessary precursors in fact. There is no natural way to produce estrogens without producing one of these androgens first. In the early phases of ovarian maturation the production of androgens dominates, so that the ratio of androgens to estrogens in early adolescent girls is high. The androgens stimulate the growth spurt along with the development of pubic and axillary hair. The small amounts of circulating estrogens are sufficient to initiate breast development. As ovarian maturation proceeds the production of estrogens by aromatization increases and the ratio of androgens to estrogens drops. Estrogen effects are more prominent in development during this phase, including the onset of menstruation, breast enlargement, subcutaneous fat deposition, and epiphyseal fusion. In boys the androgen/estrogen ratio is always high and the growth spurt is greater in magnitude.

Pathological overproduction of androgens at an early age can occur in both sexes. Often the source of excessive androgen production is not the gonads, but the adrenal gland. This condition, known as congenital adrenal hyperplasia, or CAH, is usually detected as part of routine infant screening in many hospitals and can be treated pharmacologically.

If it goes untreated, however, it can result in a precocious growth spurt and an associated advancement of skeletal age.

TURNING UP THE HEAT

Under normal circumstances it is gonadal maturation that lies behind all the various manifestations of adolescent growth and development, including the growth spurt and the development of secondary sexual characteristics. The causes of gonadal maturation are, however, not well understood. Gonadal function is controlled by the hypothalamic-pituitary-gonadal (HPG) axis, as noted previously, with the pulsatile release of GnRH by the hypothalamus allowing the production and release of gonadotropins by the pituitary, which in turn stimulate production of gametes and steroid hormones by the gonads. Gonadal steroid production in turn suppresses further production of GnRH and gonadotropins. This three-part control loop can be manipulated at any of its three anatomical nodes—hypothalamus, pituitary, or gonad—so it is perhaps not surprising that at least three general theories of the maturation of gonadal function have been advanced.

One hypothesis holds that maturation occurs in the hypothalamus. Some process that remains unclear causes the hypothalamus to begin to secrete GnRH in a regular, hourly pulse, resulting in downstream activity on the part of the pituitary and gonads. Among the primary evidence for this hypothesis is the effectiveness of artificial induction of pulsatile GnRH release in initiating gonadal maturation. In immature female rhesus monkeys the pulsatile release of exogenous GnRH directly into the hypophyseal portal system is capable of stimulating mature cyclic ovarian activity. In humans, too, pulsatile release of GnRH into the peripheral circulatory system in individuals incapable of endogenous GnRH production has been successful in provoking pubertal development. Perhaps because of the indirect route of administration, however, the results in humans are more variable than in the experimental monkeys.

The animal studies in particular suggest that the pituitary and ovary are capable of functioning before the process of reproductive maturation begins and wait only for the hypothalamus to initiate its own mature activity. Exactly what causes the hypothalamus to do this remains obscure. The endocrinologist Melvin Grumbach and his colleagues have suggested that the hypothalamus in prepubertal children is exquisitely sensitive to the negative feedback effects of gonadal steroids, and that hence only the minutest amount of testosterone or estradiol is sufficient to hold GnRH production in check. At the onset of puberty and progressively during its course the hypothalamus loses this high degree of sensitivity to negative feedback. More and more steroid is needed to hold GnRH in check. This progressive desensitization allows gonadotropin and steroid levels to rise until a new steady state is established with high circulating steroid levels indicative of mature gonadal function. Grumbach draws the analogy to a household thermostat holding the furnace's heat production in check. Reproductive maturation is analogous to resetting the thermostat to a higher ambient temperature. Fittingly, this hypothesis has become known as the "gonadostat" hypothesis.

A different hypothesis, put forward by the endocrinologist Tony Plant, suggests that it is changes in the pituitary gland, not the hypothalamus, that are responsible for the maturation of the HPG axis. Support for this hypothesis, known as the pituitary drive hypothesis, comes mainly from experimental data showing that it is not changes in GnRH production that characterize the early stages of reproductive maturation as much as increases in gonadotropin production in response to a given GnRH stimulus. Something makes the pituitary more sensitive to GnRH, this hypothesis suggests, rather than making the hypothalamus less sensitive to steroids.

A third hypothesis can be generated to be symmetrical with the other two, a "gonadal drive" hypothesis. According to this hypothesis, changes in gonadal function come first, leading the gonad to produce more steroid in response to a given level of gonadotropin stimulation. Then, over time, exposure to this higher level of gonadal steroid causes the hypothalamus to lose sensitivity to its negative feedback effects, allowing the gonadostat to reset at a higher level. (This sort of resetting of the gonadostat happens in reverse after menopause in women. Initially the cessation of ovarian estrogen production results in unrestrained gonadotropin production by the pituitary. Over time, however, under the condition of chronic low estrogen levels, gonadotropin production falls. Whether this is a result of increased hypothalamic sensitivity or reduced gonadotropin drive has not been shown.) Elevated steroid levels would also increase pituitary sensitivity to GnRH stimulation. Similar changes in pituitary sensitivity in response to estrogen levels are observed, for example, during the normal menstrual cycle.

The symmetry of these hypotheses is itself illuminating. Because the HPG axis is essentially a closed regulatory loop, resetting of the entire system can occur almost indistinguishably at any of its three anatomical levels. Whether the hypothalamus "loses" sensitivity to the suppressive effect of gonadal steroids, or the pituitary or gonad "gains" sensitivity to the stimulatory effects of GnRH or gonadotropins, respectively, the result is the same, a higher setting for the whole system. But what all three hypotheses lack is any explanation for the primary functional change, wherever it occurs. Proponents of hypothalamic desensitization suggest that the opportunity for input from outside the HPG axis is greatest at the hypothalamic level, since the brain receives information from throughout the body and from sensory organs as well. Somehow the brain must determine that the time is right for reproductive maturation and then initiate the changes in hypothalamic activity that bring it about. There is, however, the possibility that the HPG axis is affected by other inputs from outside its own domain that do not originate in the central nervous system.

One possible source of such an input is the adrenal gland. The adrenal gland shares a common embryological origin with the gonads. During the course of embryological development, the two masses of glandular tissue separate into distinct anatomical structures. The gonads become associated with duct systems that give rise to fallopian tubes or seminiferous tubules, while the adrenals become associated with clumps of postsynaptic nerve tissue. These erstwhile nerve tissues become the inner part of the adrenal gland, or adrenal medulla. This part is responsible for secreting epinepherine and norepinephrine into the general circulation, where they act as hormones. The outer part of the adrenal gland, the adrenal cortex, retains the enzyme systems necessary for the production of steroid hormones, essentially the same system as possessed by the gonads.

The adrenal gland, however, specializes in the production of two steroids not produced by the gonads, cortisol and aldosterone. Cortisol, as we have seen, plays an important role in energy metabolism, and because of that central function also plays a part in parturition and lactation. Aldosterone helps to regulate mineral balance as part of a system controlling kidney function. Perhaps it is this important function that accounts for the anatomical location of the adrenal glands on top of the kidneys. In addition to cortisol and aldosterone, the adrenal gland also produces androgens, not testosterone but the so-called adrenal androgens dihydroepiandrosterone (DHEA) and androstenedione. Although weaker in their biological effects than testosterone, both of these hormones are capable of binding to androgen receptors. They can, in addition, be converted by enzymes occurring in various target tissues into more potent steroids, both androgens and estrogens. Production of adrenal androgens occurs at rather high levels in the fetus prior to birth, but ceases soon after. Late in childhood, around age six to eight in girls, eight to ten in boys, production of adrenal androgens starts up again. The increase in adrenal androgens, sometimes termed adrenarche, precedes the increases in gonadal steroids in both sexes and may contribute to the adolescent growth spurt in girls. It is also possible that the increases in adrenal androgen levels circulating in the blood play a role in resetting the HPG axis, either by desensitizing the hypothalamus or by increasing the sensitivity of the pituitary or gonads to their respective upstream signals.

The ability of adrenal androgens to produce such an effect is illustrated by cases of untreated CAH. In these cases the adrenal gland produces inappropriately high levels of androgens from birth. Not only do those who suffer from this condition undergo premature growth spurts, but they also undergo premature reproductive development to the point of mature gonadal production of gametes and hormones. Girls menstruate and ovulate and boys produce viable sperm. Bringing the adrenal androgens under control after this point does not reverse the reproductive maturation that has already been achieved any more than it reverses the growth spurt that has already occurred. But just because adrenal androgens *can* have this effect in pathological cases doesn't prove that they ordinarily *do* have this effect. On the negative side, there are cases of delayed or absent adrenarche. Children with this condition usually undergo a normal, if delayed, process of reproductive maturation.

But even if adrenal androgens do turn out to play a role in the normal maturation of the HPG axis, this only pushes the question one step back. What, we would then ask, causes adrenarche? Physiological causation often has this tendency to appear as a never-ending chain of proximate causes receding into the distance. But none of the mechanisms that have been described offers an adequate explanation for the timing of adolescent maturation. Why does it happen when it does? Or, to revisit the question with which this chapter opened, why does it take so long for it to start? Why the long drawn-out process of childhood growth before the final physiological frenzy of the adolescent growth spurt and reproductive maturation?

THE THRESHOLD OF MATURITY

In 1970 Rose Frisch and her colleague Roger Revelle at Harvard advanced a radical approach to the question of the timing of adolescent maturation. The hypothesis they advanced, sometimes referred to as the Frisch hypothesis, metamorphosed over the succeeding years into different specific forms, but sustained at its core the novel idea that reproductive maturation was primarily synchronized with the growth of soft tissue rather than the skeleton. Latent in this proposal was an evolutionary explanation for the timing of adolescent maturation that was not clearly articulated until later. The Frisch hypothesis has been subjected to withering criticism that is wholly convincing to those who take the time to consider the details. But, like the nursing frequency hypothesis, the Frisch hypothesis has proven so intuitively appealing and comes so close to the truth that it continues to be widely accepted.

In the late 1960s Rose Frisch and Roger Revelle were studying the relationship of nutrition to fertility and population growth. The Malthusian paradigm, which then dominated the thinking of demographers and other social scientists, made this a suspect research agenda. But Roger Revelle, together with others such as the Stanford scientist Paul Ehrlich, was motivated by a new concern for the ecological consequences of human population growth. These new human ecologists tended to adopt the familiar framework of systems ecology, tracing the flow of energy and limiting resources through the human population and its activities. It was a natural part of this systems approach to consider how energy inputs would affect the output of additional people. Data on which to base modeling efforts and other conclusions were difficult to come by, however, and artificial control of fertility in developed countries complicated the analysis. The most promising result of Frisch and Revelle's early efforts came from analyzing the relationship between energy intake and the timing of adolescent maturation in different populations. Where data on physical growth were available, they observed weak negative correlations between the age at peak adolescent growth and average caloric intake. On a population level, however, it was difficult to find more specific data with which to probe the relationship.

Frisch and Revelle found richer data finally in the longitudinal growth studies that had been performed in the United States during the earlier part of the century. Stimulated in large part by the ideas of Franz Boas, researchers had followed and measured several cohorts of children regularly from childhood through adolescence. Similar studies were carried out in the United Kingdom and other countries. These studies had allowed for the development of standards for the staging of secondary sexual development (Tanner stages) and skeletal development as well as height, weight, and other physical measurements. As a body these studies provided the empirical basis for the conclusion that reproductive maturation was closely synchronized with skeletal maturation. For Frisch and Revelle they provided an opportunity to look below the level of aggregate population statistics to the growth and maturation histories of individuals. It was at this level that they turned the conventional wisdom on its head.

Frisch and Revelle combined data from three longitudinal growth studies conducted in Berkeley, California, Denver, Colorado, and Boston, Massachusetts. These data sets did not include information on caloric intake, but they did contain information on one important index of reproductive maturation in girls: the age at menarche. They examined the relationship between this event and growth in weight and height. They found, as had others before them, that girls who reached menarche later tended to be taller at menarche than girls who reached menarche sooner. In contrast, however, they found that the average weights of early and late maturing girls at menarche were virtually the same: approximately forty-seven kilograms. They reported their finding in the prestigious journal *Science* in 1970 together with a radical interpretation. It is not growth in height that is coordinated with reproductive maturation, they asserted, but growth in weight. They hypothesized that forty-seven kilograms represents a "critical weight" that "triggers" menarche in girls. Information that this "critical metabolic mass" has been attained was somehow communicated to the hypothalamus, where it caused the changes that initiate mature HPG activity. This mechanism would account for the effect of nutrition on the timing of reproductive maturation since well-nourished individuals would reach the critical weight earlier. It could even account for the observed historical decline in age at menarche in Europe and the United States, a phenomenon known as the secular trend. In 1972 Frisch used historical data from Scandinavia to show that the average weight at menarche had remained constant through time at about forty-seven kilograms during the secular trend while the average height at menarche had increased. Although their data were limited to girls where a convenient marker of reproductive maturation existed in the data sets available, Frisch and Revelle suggested that similar mechanisms might regulate reproductive maturation in boys.

The data were presented in the *Science* article in the form of average

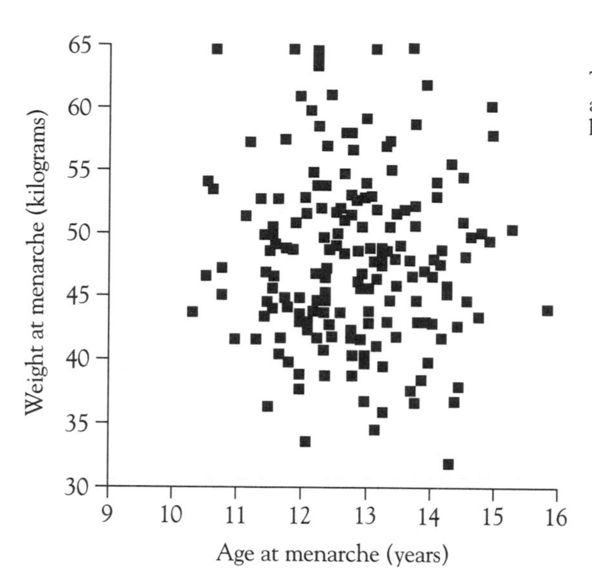

The original data for Frisch and Revelle's critical weight hypothesis.

values of weight and height for different categories of menarcheal age (before age twelve, twelve to thirteen, thirteen to fourteen, and fourteen to fifteen). The next year Frisch and Revelle published a more extensive report of their analyses in the British journal *Archives of Disease in Childhood.* In this more obscure publication appears a revealing plot of all the individual data, a figure that instantly makes clear the central defect of the first version of the Frisch hypothesis. The plot of weight at menarche versus age at menarche appears as a cloud of points with tremendous variability on the weight axis. Some girls reach menarche at weights as low as thirty kilograms or as high as sixty-five kilograms, and between these extremes there appears to be only random scatter. In fact, the graph (shown above) is a textbook example of a zero correlation: two variables that have no relationship to each other whatsoever. The constant average weight at menarche in this case is entirely spurious. The same statement could be made about the average of the last two digits of the license plate on the family car and the age at menarche of the oldest daughter. Because there is no relationship between these variables, the average license plate number will be fifty for early maturers, late maturers, and every group in between.

This egregious misinterpretation of the data was soon pointed out. Attention was drawn to the enormous variability in weight at menarche within a population and the incompatibility of this fact with a hypothesis maintaining that a critical weight triggers menarche. In addition, Francis Johnston, an anthropologist, and his colleagues pointed out that average weight at menarche varies widely between populations. In general, girls who are tall for their age in childhood tend to mature earlier than others do. Only within height categories is it true that heavier girls mature earlier; only those who are heavy for their height mature earlier, as opposed to those who are heavy absolutely.

In response to these criticisms, Frisch and Revelle joined with the statistician Robin Cook to revise their hypothesis. Perhaps taking their lead from the observations of Johnston and his colleagues, they shifted their attention to weight for height rather than simple weight. Weight for height, they reasoned, was an indirect measure of body composition, a way to distinguish people who are heavy because they are fat from people who are heavy because they are tall. Perhaps a better measure could be found. They proposed using indirect estimates of body composition, derived from weight and height, instead. Because living cells are primarily composed of water, people can be modeled as if they were composed of three simple elements: bone, fat, and water. Various methods exist for estimating the percentage of these components based on buoyancy in water, resistance to the passage of electric current, or the dilution of trace chemicals as they pass through the body. Frisch and her colleagues made use of a set of equations generated to relate estimates of body composition, in particular total body water, to measures of height and weight. When they plotted the percentage of total body water derived in this way against age at menarche for their original set of data they made a satisfying observation: the variance in percentage body water at menarche was much less than the variance in weight had been. Rather than falling into a random scatter of points, the data now appeared to cluster more tightly along a line representing a constant body composition at menarche. Frisch and her colleagues published this result together with a new version of the Frisch hypothesis: rather than a critical weight, there is a critical body composition that triggers menarche in girls. Because girls accumulate fat during adolescence rather than lose it, they interpreted the critical body composition as a critical level of fat.

It did not take long for a fatal flaw in this analysis to be pointed out. The reduction in variance that had been achieved by switching the "critical" variable from weight to fatness was spurious. It was a result of using an *estimate* of body composition rather than a measurement: The procedure Frisch and her colleagues had used assumed that all girls of the same weight and height have exactly the same body composition. Of course, this isn't true. One girl may have more fat, another more muscle, and yet be the same weight and height. The estimate Frisch and colleagues had used came from regression equations relating height and weight to an average amount of total body water. But those same equations also included an error term representing the variance in actual body water to be expected around that average. The reduction in variance that had been achieved in the new analysis was exactly accounted for by the variance that had been thrown away by neglecting this error term. No magic had been performed by switching variables. The data still fell far short of providing any support for a critical level of weight or fatness in triggering menarche.

Uncowed by these criticisms, Frisch put her hypothesis through yet another transformation in a paper published with Janet McArthur in *Science* in 1974. Here the same data from the same three longitudinal growth studies are presented graphically yet again, but this time as a plot of weight at menarche against height at menarche. Diagonal lines drawn across the plot represent centiles of weight for height. Using the same regression equations as before, Frisch and McArthur interpret these lines as centiles of estimated fatness. This time, rather than arguing for the significance of the average value of weight or fatness, Frisch and McArthur draw attention to the lower end of the distribution of weight for height. Pointing out that very few girls in the sample reach menarche below a line representing 17 percent of body weight as fat, they hypothesize that this level of fatness may be a "minimum threshold" for menarche to occur. Anticipating a recurrence of the familiar objections based on the large variance in fatness at menarche, they explicitly state that this hypothesis does not account for variance in fatness above the minimum threshold. Rather, they suggest that 17 percent fat represents a necessary but not sufficient condition for the attainment of menarche.

In the same paper Frisch and McArthur present an analysis of data from adult anorexia nervosa patients. Anorexia nervosa is a psychological condition that features self-imposed quasi-starvation as one of its principal symptoms. Sufferers are often amenorrheic. Successful treatment results in increasing food intake, weight gain, and often resumption of menstruation. In the 1974 article Frisch and McArthur presented data on a group of women treated for anorexia nervosa associated with amenorrhea. The data are plotted to represent weight and height before and after resumption of menstruation. As with the data on the menarcheal age girls, diagonal lines on the graph represent different levels of body fatness. All the points representing the resumption of menses fall above the line representing 22 percent fat, while many of the points representing the amenorrheic state fall below it. On this basis Frisch and McArthur broaden the hypothesis to include a minimum fatness (22 percent) necessary for the maintenance of menstruation in adulthood in addition to a (different) minimum (17 percent) necessary for its onset in adolescence.

Despite the effort to ward off criticism based on the high variance in weight for height at menarche, the new "minimum fatness" version of the Frisch hypothesis remained vulnerable to censure. W. Z. Billewicz and his colleagues pointed out that for a fatness threshold to be biologically meaningful the probability of reaching menarche in the next unit of time would have to be greater after it is crossed than before. Otherwise there is no evidence that attainment of the threshold has any relationship to the attainment of menarche. After all, one could identify a minimum shoe size below which menarche is unlikely to occur, or minimum values on any number of scales that have no particular relationship to menarche. Longitudinal growth data, however, show no such effect of the 17 percent fatness threshold. When one controls for the effect of advancing age, the probability of reaching menarche in the next year is no greater for girls over the 17 percent fat threshold than those below it.

Noël Cameron, of the Institute of Child Health in London, made the same point in a different way. If there were a threshold value of fatness that raised the probability of menarche, then the variance in fatness of girls at menarche should be less than the variance in fatness of girls one year or two years before menarche. After all, some girls will grow quickly and approach the fatness threshold rapidly, gaining a lot of fat over the year or two before the event. Other girls will grow slowly and gain only a little fat over the same period. Two years before menarche, then, there should be more variance between the fast and slow growers than at menarche. Unfortunately, there isn't. The variance in fatness among girls increases as menarche is approached and passed, rather than decreases, in both British and American data sets.

Through all its avatars the Frisch hypothesis has been subject to a central problem: it simply isn't supported by the data, even the data that were used to generate it. No relationship can be demonstrated between the attainment of any particular weight or fatness and the probability of menarche occurring. The variances of weight and fatness at menarche are simply too large and the correlations in time between growth on these dimensions and the onset of menstruation too weak. Some correlation will exist, of course, between any pair of variables that themselves reflect an underlying developmental process without the necessity of any causal connection between them. Consider the depth of the tire tread on a car and the first failure of one of the headlight bulbs. The greater the tread wear, the more likely the headlight is to fail. There is probably even a minimum tread wear below which headlight failure is unlikely. But that doesn't mean that headlight failure is *caused* in even an indirect way by tread wear. Rather, both are functions of the mileage on the odometer. In the same way, changes in body composition and the onset of menses may evidence some correlation by virtue of being reflections of a common process of maturation. There is no empirical evidence that the relationship is any tighter than that. In fact, as noted previously, increases in fatness in adolescent girls occur in response to production of estrogen by the ovary, the same estrogen that causes endometrial development and menstruation. Thus rather than increases in fatness "triggering" menarche, increasing estrogen production causes both menarche and increases in fatness. It is the increase in estrogen production itself that represents reproductive maturation; it is that increase that must be explained. And the increasing production of estrogen cannot be explained by one of its own effects, an increase in fatness.

SIZE MATTERS

The empirical critiques of the Frisch hypothesis in its various forms remain valid today. The hypothesis has changed shape in an effort to answer them, but without success. Yet the hypothesis itself is still alive and well and continues to be cited in textbooks, scientific papers, and popular accounts as true. Part of the reason for this curious situation may be that the critiques are presented in statistical terms that require a familiarity with the data in question and with methods appropriate for analyzing them. Against these statistical arguments stands an elegantly simple hypothesis that many people feel "makes sense." The sense they see in it, however, lies not in any physiological mechanism or statistical demonstration, but in an associated assertion of function that is fundamentally evolutionary. Girls need to accumulate fat, Frisch argues, in order to reproduce successfully. Until they have accumulated enough fat to render the probability of successful reproduction sufficiently high, reproductive maturation would be untimely and reproduction itself a waste of effort. Natural selection, therefore, should have designed the reproductive system to wait until a sufficient reserve of fat has been accumulated before initiating mature reproductive function.

As a functional explanation this argument has great appeal. It explains (or seems to) why women are fatter than men. It explains (or seems to) why anorexic women don't menstruate. It explains (or seems to) why better nutrition leads to earlier maturation. The empirical critiques seem like so many statistical quibbles compared to the commanding explanatory power of the Frisch hypothesis and its functional interpretation. But there is one question it doesn't satisfactorily answer, the question that opened this chapter. Why does it take so long for reproductive maturation to occur? If attaining a minimum level of fatness is the effective constraint on reproductive success, why waste all that time and energy on skeletal growth? Why not stop growing skeletally at age eight or age six or age three, put on fat instead, and get a decade's jump on everyone else in the evolutionary horse race? All other things being equal, this physiological strategy would have a huge selective advantage over slow maturation. In short, why grow "up"?

Attractive hypotheses, like the nursing frequency hypothesis and the Frisch hypothesis, often owe part of their attraction to the absence of alternatives. They seem to explain phenomena for which no other explanation exists. But almost any explanation wins out when tested against no explanation. Science advances most effectively by testing alternative explanations against each other, to determine not which is "true," but which is "better." The nursing frequency hypothesis is best tested not against the hypothesis that lactation has no effect on ovarian function, but against the relative metabolic load hypothesis. Similarly, the Frisch hypothesis is best tested not against the hypothesis that there is no explanation for the synchronization of physical growth and reproductive maturation, but against alternative explanations for that synchronization.

One alternative, presented by Kim Hill and Magdalena Hurtado of the University of New Mexico, bears a surface similarity to the Frisch hypothesis. Their alternative is derived from the branch of evolutionary ecology known as life history theory and owes a particular debt to the ideas of the University of Utah ecologist Eric Charnov. Briefly, Hill and Hurtado start with the fundamental proposition that organisms must distribute metabolic resources between competing physiological categories of maintenance, growth, and reproduction. The amount of energy available for growth in childhood, or reproduction in adulthood, is determined by the total amount of metabolic energy the organism produces minus the amount invested in maintenance. Reproductive maturation represents a shift in energy allocation from growth to reproduction, a shift that in humans and many other mammals is made rather abruptly. The timing of this shift, Hill and Hurtado contend, is dictated by the relative benefit of continued growth versus the benefit of initiating reproduction. In general, the larger an organism gets, the more residual energy it will have to invest in reproduction and so the greater its effective fertility. The increase in fertility that can be achieved with further growth appears to decrease with size, however, both in theory and according to available data. Putting off reproduction, on the other hand, always has a fitness cost. There is an increased risk of death before beginning to reproduce as well as an increase in generation length, both of which lower overall fitness. At some point the gain in fertility that can be achieved by further growth is less than the fitness cost of further delay. That is the optimal point at which to stop growing and begin reproducing. This approach works quite well, Hill and Hurtado show, in predicting the final size and average age at first reproduction of hunter-gatherer populations like the Ache of Paraguay and the !Kung of Botswana.

In some ways this hypothesis seems to resemble the Frisch hypothesis in its emphasis on size and the energetic cost of reproduction. There are important differences, however, that should be noted. The Frisch hypothesis stresses the accumulation of stored energy as crucial to successful reproduction, particularly stressing the accumulation of fat. Size per se does not figure in the hypothesis. There is nothing in the hypothesis to explain why an individual should wait to achieve a certain size before accumulating the stored fat deemed necessary for reproductive success. Hill and Hurtado's hypothesis, by contrast, is explicitly linked to size at maturity through the relationship of size to metabolic production, maintenance requirements, and hence the residual available for reproduction. Frisch's hypothesis stresses a status variable, how much stored energy has accumulated. Hill and Hurtado's hypothesis stresses a dynamic variable, how much residual energy is being produced.

Yet despite its emphasis on size, Hill and Hurtado's hypothesis also shares some of the same weaknesses as the Frisch hypothesis. It doesn't really explain why growth in *height* is so important as opposed to growth in soft tissue. Metabolic rate, maintenance requirements, and residual energy available for growth and reproduction tend to scale with mass across species. A shorter, stockier version of an organism can have a higher metabolic production than a taller, more slender version. Hill and Hurtado can use their approach to generate predictions about adult size in terms of height by making use of the standard relationship of weight and height in humans. Such-and-such an average mass implies such-and-such an average height. Conversely, they can use the empirical weight-height relationship to generate predictions of metabolic production for a given height. But in this way they are assuming part of the answer. Why aren't humans shorter and stockier? Why can't the same metabolic mass be carried on a smaller skeleton? Why afford skeletal growth a higher metabolic priority than soft tissue growth during childhood? Why isn't growing "out" just as good as growing "up?" In addition, although the Hill and Hurtado hypothesis successfully predicts average size and age at reproductive maturity for populations, it is not clear whether it is equally successful in explaining variation between individuals.

Skeletal growth, on the other hand, meets the conditions one would expect of the key variable in a threshold hypothesis better than growth in weight or change in body composition. For example, it was noted above that if there were a threshold value of fatness that made menarche more likely, variance in fatness should be less at menarche and greater at increasing intervals of time before menarche. While neither weight, weight-for-height, nor body composition meets this condition, height does. Although late maturers tend to be taller at menarche than early maturers, the difference in their heights at menarche is far less that it is a year or two before menarche. By age eighteen, when growth is virtually ended and reproductive maturation is more complete than at menarche, early and late maturers have converged on essentially the same average height while their average weights have continued to diverge.

Frisch has claimed that the timing of menarche in girls can be predicted from their relative weight and estimated body fatness in late childhood. In the journal *Pediatrics* she produced a graphical method that physicians could use to make such predictions. The accuracy of the method was not tested against an independent sample, different from the one used to generate the prediction equations. But the error terms of the prediction equations themselves suggest that the method, while statistically significant, is not particularly accurate. Predictions based on the rate of growth in height are in fact much more accurate even when tested against an independent sample. Special techniques can also be used to statistically disentangle growth in height from growth in weight in the same sample, allowing the strength of their respective correlations with menarcheal age to be compared. In such an analysis growth in height has a much stronger independent correlation than growth in weight or weight for height.

Empirically, menarche is much more closely tied to the attainment of a threshold height than a threshold weight or body composition. What seems to be missing, though, is a functional understanding of why this is so. Why should height be so important in determining a woman's ability to reproduce successfully? One possibility is that height itself is not the key variable, but only a closely linked one. If we ask what aspect of skeletal size other than height might impose a limiting constraint on reproductive ability, the most likely answer is: the pelvis. Why isn't a six-year-old girl able to reproduce successfully? Not because she lacks fat, but because the baby would never make it through her birth canal. There are, in fact, rare cases on record of girls with pathological precocious puberty giving birth as early as age four, but only by Caesarean section.

The interior dimensions of the pelvis are difficult to measure directly. But there are proxy measurements that might serve even better than height. Biiliac diameter is a measurement of the breadth of the pelvis at the top of the iliac crests, the bony protuberances usually found on each side of your stomach under your belt. Although it is not a measurement of the birth canal itself, which is lower down in the pelvis, it does provide an index of overall pelvic size. Biiliac diameter is also reported in some of the available longitudinal growth data sets. The average value of this index for adult females is about twenty-seven centimeters. If we "try on" a biiliac diameter slightly smaller than this, twenty-four or twenty-five centimeters, as a "pelvic size threshold" for menarche, we find a remarkably good fit. The age at which this threshold diameter is attained is strongly correlated with the age at menarche. Those who pass it earlier reach menarche earlier while those who pass it later reach menarche later, and nearly all girls tend to reach menarche within a year or so of passing the threshold. This is exactly the sort of relationship that could not be demonstrated for fatness. If we broaden our perspective to a comparison of populations, the same relationship holds. Among twenty-four populations for which the average age at the attainment of a biiliac diameter of twenty-four centimeters and the average age at menarche are both available, the correlation between these two ages is very high (0.8), with essentially the same slope as that found in the data for individuals within a population.

Longitudinal growth data on internal pelvic dimensions are understandably harder to come by. Nowadays, ultrasound provides a practical method for making the requisite measurements. But at the time that most of the existing longitudinal growth studies were conducted, the only way to collect these data was from x-rays. Pelvic x-rays expose the subject to considerable radiation focused directly on the area containing the gonads, maximizing the risk of genetic mutations in the germ cells themselves. Performing these measurements on preadolescent girls who have their entire reproductive lives ahead of them entails the greatest risk of harm. Nevertheless, at least one longitudinal growth study included pelvic x-rays, from which the anthropologist Marquisa Moerman was later able to estimate critical internal diameters of the birth canal. This series of x-rays demonstrates the dramatic reshaping of the female pelvis that occurs during adolescence, with increases in all the critical diameters of the birth canal. As noted in a previous chapter, spreading of the hip joints is also associated with a wider flaring of the iliac blades to maintain an efficient angle of action for the hip muscles. For this reason internal pelvic growth is correlated with growth in the biiliac diameter. All of this reshaping occurs at the very end of skeletal growth under the influence of the same estrogen hormones that bring growth of the long bones to a halt. Pelvic reshaping is thus designed to occur only when final skeletal size has been attained. As this final phase of skeletal maturation takes place, early and late maturers converge on the same internal pelvic diameters, diameters that are, as we have seen, big enough (but only just!) for the birth of an infant human head.

The "pelvic size" hypothesis thus stands as a strong alternative to the Frisch hypothesis and the Hill-Hurtado hypothesis. Empirically, the pelvic size hypothesis has significant advantages over the others in its ability to predict the timing of reproductive maturation both in individuals and in populations. It also accords with, rather than conflicting with, the evidence of a higher metabolic priority on skeletal growth than on soft tissue growth during childhood. Unlike the other two hypotheses, the pelvic size hypothesis is based on the notion that there are mechanical constraints on successful reproduction as well as energetic ones. But the mechanical constraints are viewed as prior. Until a girl is large enough to give birth, it doesn't matter whether she has enough energy to grow a baby. Unlike the other two alternatives, the pelvic size hypothesis alone offers an explanation of why so much time is devoted to growing "up."

Or does it? Don't hypothetical alternatives still exist? Why not stop growing earlier and have a baby with a smaller head? As discussed earlier, it appears that a large head is an important part of being human, and that gestation is continued as long as possible given the high energy demand of the developing fetal brain. Earlier reproductive maturation presumably was possible when pelvic size wasn't a constraint on parturition. Other constraints, energetic ones included, may then have been more important. Why not grow the pelvis disproportionately so that it can attain the critical dimension while the rest of the skeleton is still small? Perhaps the pelvis is growing as fast as it can already, though in the absence of supporting data that is a weak argument. More compelling, perhaps, are arguments based on the mechanical trade-offs with bipedal locomotion noted previously. A disproportionately large pelvis might not be compatible with an ecology that requires extensive bipedal walking and carrying, or with other physical tasks important to making a living as a human. The selective pressure for increased brain development in a species that was already committed to bipedalism effectively made skeletal size in general and pelvic size in particular novel constraints on female reproductive ability and hence imposed a new threshold for reproductive maturation.

Once the mechanical constraints of physical size have been overcome, energetic constraints may indeed become limiting on female reproductive success. We will see in the next chapter that there is considerable information on this point. The pelvic size hypothesis does not preclude the importance of energetic constraints on adult female reproductive function. Rather it suggests that the size constraint must be overcome first, and to that extent skeletal growth will enjoy a priority over energy storage until a mature size has been achieved. This prediction is also borne out by empirical evidence. As noted, the acceleration of fat accumulation by girls is a consequence of ovarian estrogen production, not an antecedent or a cause. When a mature size is achieved, pelvic remodeling occurs and energy allocation is diverted away from skeletal growth toward growth in weight and fat storage. The sequence of maturational events corresponds to the logical ordering of priorities suggested by the pelvic size hypothesis.

All three hypotheses share some elements in common. All three, for instance, allow childhood nutrition to play an important role in the timing of reproductive maturation. All three share the limitation, however, of focusing on female maturation. In part this reflects the fact that constraints on female reproductive success are easier to identify and "markers" of female reproductive maturation, like menarche, are easier to record. Frisch has implied in some of her writings that male maturation and adult fecundity are affected by nutrition in a manner broadly analogous to the effects in females. Certainly the secular trend in growth that has occurred in many populations has affected both sexes. Given the relationship of the adolescent growth spurt in males to increasing testicular hormone production, it is reasonable to assume a concurrent advancement of reproductive maturation. The specifics of the Frisch hypothesis, however, do not logically apply to males. Male reproduction does not require the same physiological investment of energy that female reproduction does, nor do boys accumulate the proportionate amounts of fat that girls do either before or after puberty.

The Hill-Hurtado hypothesis can formally be applied to males without modification on the basis of the empirical relationship between size and reproductive success in men. Without specifying exactly how metabolic energy is converted into reproductive success, the Hill-Hurtado hypothesis can simply be based on the black-box assumption that it must be. Empirical evidence on the number of surviving offspring that men of different size have can be used to estimate the net benefit that would accrue to prolonging growth versus ending growth and beginning to reproduce. When that net benefit drops below zero, the optimal time for reproductive maturation has arrived. Fitting Ache data to such a model indicates that Ache men stop growing at approximately the predicted size. The basic structure of the hypothesis and its life historical approach is sound, but the specifics are largely missing. How does metabolic energy translate to reproductive success for males? Does it begin to do so as soon as they stop growing, or only when they begin to reproduce (which may be many years later)? Does the model help us understand individual variation in the timing of reproductive maturation, or does it apply only to population averages? Does it even help us explain those populations averages satisfactorily without additional assumptions? Why do males mature later than females, for instance? Saying that they have not yet reached the optimal size in terms of reproductive payoffs at the age that girls reach theirs may be an accurate empirical description, but is it an adequate explanation?

The pelvic size hypothesis, of course, does not apply to males. Not only are men free of the energetic constraints of pregnancy and lactation, they are free of the mechanical constraints of giving birth. These differences in reproductive constraints between men and women are reflected by some of the statistically significant physical dimorphisms our species displays, the differences in pelvic morphology and body composition. But males still continue to grow "up" for years before maturing reproductively. Some reproductive advantage must derive from the increased size or they would stop growing earlier. The prolonged period of male growth can't simply be due to a developmental constraint that "forces" them into the same pattern as females either, since they in fact mature on a different schedule.

Some clues may lie in the pattern of male physical growth in relation to maturation. For instance, the dimorphism in body composition that develops at puberty in humans is not only a consequence of increases in fatness on the part of females, but is also a consequence of increases in muscle mass on the part of males. In the United States, for example, boys and girls are quite similar in terms of body composition at the end of childhood. Physical measurements of subcutaneous fat and arm muscle area reflect this. At age ten, the fiftieth centile of combined arm and back subcutaneous fat thickness in girls is approximately 20 millimeters, versus 15.5 millimeters in boys. The fiftieth centile of cross-sectional arm muscle area at the same age is 23.5 square centimeters for girls, 25.5 square centimeters for boys. By age seventeen the fiftieth centile for fat thickness has increased to 31 millimeters in girls but only to 16 millimeters in boys. The fiftieth centile of arm muscle area in girls increases to about 33 square centimeters by age seventeen, but in boys has surpassed 53 square centimeters. Related to this is the striking divergence in strength between the sexes. In one U.S. longitudinal growth study both boys and girls have similar average grip strength at age ten (about 70 kilograms). By age eighteen, however, mean grip strength in males has increased to over 210 kilograms while the mean for females has risen only to about 120 kilograms. Males and females are clearly doing different things with residual metabolic energy after growth in stature ends. These different allocations both presumably reflect investment toward reproduction, however.

Another clue may be the higher heritability of biacromial diameter, or the breadth of the shoulders, in males than in females. The higher heritability of biiliac diameter in females can be viewed as consistent with the greater importance of this skeletal dimension for females' fitness, leading to its relatively greater genetic control and relatively lower sensitivity to environmental effects. Something similar appears to be true for shoulder breadth in males, indicating that this aspect of male physique may be relatively important to male fitness.

These clues suggest that an effective hypothesis regarding the timing of male maturation may proceed from a better understanding of male reproductive success and its constraints. We will therefore revisit this question when we consider adult male reproductive function in greater detail in a later chapter. First, however, we will turn to the regulation of adult female reproductive function once it has been freed of the constraints of physical size.

BALANCING ACT

IN GENERAL, there are two major constraints on a female mammal's ultimate reproductive success: energy and time. The metabolic task of converting energy from the environment into viable offspring falls to the female, and the rate at which she can produce offspring is limited by the rate at which she can direct metabolic energy to the task. Available metabolic energy can clearly be limiting on this process, but for females a second limiting resource is time. In humans, for example, a minimum of nine months of gestation and some period of intensive lactation must ordinarily be devoted to each surviving offspring. This, together with a finite reproductive span, puts an upper limit on the number of offspring a woman can produce in a lifetime. Viable offspring represent relatively inflexible quanta of investment, both in terms of energy and in terms of time. Half a baby or half a gestation is worse than none. The product is not viable, and the time and energy invested is lost. Nor does it appear to be an option to grow a baby at half the normal rate and spread the burden of gestation over eighteen months. Viewed naively, it might seem that such a strategy would be a useful response to a shortage of metabolic energy. Much of human development, however, appears to be constrained to a relatively fixed timetable so that this sort of physiological flexibility is not available. In natural fertility populations therefore, and throughout the majority of human evolution, making up for time lost early in a reproductive career by accelerating the rate of reproduction later has not been much of an option for females. In contrast, male reproductive success is not subject to the constraints of time and energy in the same degree. Male mammals may, at least potentially, inseminate many females within short periods of time without having to assume the metabolic burden of all the resulting gestations. A male may not begin to reproduce until relatively late in life and still leave more offspring than a contemporary who begins to reproduce at a young age. Hence neither time nor metabolic energy places the same severe constraints on male reproductive performance.

Female reproductive success may be subject to other limitations and constraints, of course. Deficiencies of other nutrients besides energy, for example, may have a negative effect on human fetal growth and the viability of offspring. In general, however, these constraints are encountered much less frequently than energy deficiencies. Particularly in natural fertility populations, as noted earlier, prematurity and low birth weight are the primary risks to infant survival, both of which are most often consequences of low energy availability. Males may help females meet the metabolic load of gestation and lactation by provisioning them with food. Courtship and mating rituals in many species involve stereotypical displays of provisioning behavior or physical traits that are correlated with provisioning ability. In humans, cross-cultural studies of mating preference consistently indicate that women, on average, prefer male traits such as wealth and socioeconomic status that predict provisioning ability over traits such as youth or physical attractiveness.

Female reproductive success can also be limited by the genetic quality of her mate, since her offspring will receive half their genes from their father. Females of many species appear to discriminate between males on the basis of phenotypic traits other than those associated with provisioning, traits that may indicate genetic quality in various ways. Humans, for example, appear to place a great deal of weight, subconsciously perhaps, on physical symmetry, especially in the face, in judging attractiveness. Such symmetry, it has been argued, represents a more complete buffering of development from random environmental effects, and hence may be an important marker of genetic quality. Offspring fitness is not only a matter of paternal genes, of course. The quality of the maternal genetic contribution is also important. We have already considered the fact that early pregnancy loss may effectively discriminate embryos on the basis of their genetic quality. These factors are secondary, however, to the constraints of energy and time for females. If there is not enough energy or time to produce a viable neonate, it doesn't matter what genes it is carrying.

It can be presumed that natural selection has shaped female reproductive physiology in humans, as in other mammals, to make the most effective use of time and energy in producing offspring. To a large extent these amount to the same thing, the optimal spacing of births. As we have seen, optimal birth spacing in turn involves physiological "decisions" over the allocation of energy between competing categories: continued milk production, maternal maintenance costs, a new pregnancy. Designing female reproductive physiology to be sensitive to the relative metabolic load of lactation results in an elegant balance among these three. The metabolic competition between consecutive offspring is minimized, along with the probability of the mother's having to metabolize for three. Anything that hastens the transition of an offspring from metabolic dependence on the mother (such as the availability of suitable weaning foods) hastens the mother's resumption of fecundity. Anything that increases the available metabolic energy above the mother's maintenance requirements or reduces her need to cut metabolic corners related to her own maintenance also hastens her resumption of fecundity. But even when investment in lactation has diminished to insignificance, a physiological decision must still be made. This allocation decision can be thought of as involving three basic categories, investment in a pregnancy, investment in maternal maintenance, or investment in energy storage. Energy can be used for reproduction or maintenance now, or it can be stored for allocation to one or both of those categories at a later time.

RAINY DAY PHYSIOLOGY

Humans have a remarkable capacity to make use of energy storage. Our ability to store fat in subcutaneous adipose deposits is particularly well developed. Few mammals carry as much fat under their skins as we do unless they are preparing to hibernate or have become largely aquatic. We appear to exceed most of our primate relatives in this regard, with the possible exception of the orangutan. Because this subcutaneous fat functions as insulation as well, we have also lost the covering of body hair that other primates enjoy, even those living in equatorial regions and savannas. Other than pubic and axillary hair (which serves other functions) and the facial and body hair that some males display (which probably results from sexual selection), we retain only a rather bizarre skullcap of hair in both sexes. The skull, of course, represents a portion of the anatomy where a great deal of metabolic heat is both produced and potentially lost, and where very little subcutaneous fat deposition occurs. But our subcutaneous fat does not primarily serve the function of insulation. Indeed, it appears to generate more of a thermoregulatory difficulty than a solution. When metabolic heat needs to be dissipated we must shunt blood to alternative venous passages located near our skin surface above the layers of subcutaneous fat on our limbs, face, and trunk. Even then our limited capacity for radiative heat loss forces us to rely on evaporative cooling to maintain our body temperature within tolerable limits.

Rather than insulation, fat represents a reservoir of energy that is mobilizable in time of need. Females may carry more fat on their bodies in proportion to their weight, but in absolute terms males and females carry comparable reserves of calories. Fat is a denser form of energy storage than the glycogen that we store in our liver and muscles. Glycogen serves as a shorter-term, more immediate form of mobilizable energy, the energy we call upon to perform demanding physical tasks. Fat represents a qualitatively different form of energy storage. Metabolic energy must be spent to form fat for storage and again to mobilize the fat later. But these metabolic "taxes" on fat creation and mobilization are offset by the fact that fat, once formed, is metabolically inert. The only direct cost it entails is the extra weight it represents. Even that is minimized by its high caloric content for its molecular weight and its hydrophobic nature. The reduced water content of fat cells makes them lighter for their volume than other cells. There are other differences between glycogen and fat that make glycogen more versatile as a metabolic substrate. Glycogen can be used anaerobically, for example, to fuel quick bursts of muscular activity, or to support energy expenditure above the upper limit of oxygen delivery for short periods. Fat ordinarily serves only as a substrate for processes that generate energy by using oxygen and are thus limited by the rate at which oxygen can be delivered to and utilized by cells.

But if fat can't give you the same energy surge that a candy bar can, it can keep you alive for a lot longer. For their size and weight, humans can survive for an exceptionally long time on little or no energy consumption. During famines or self-imposed starvation, humans have been known to live for six months or more as long as water is available. Muscle tissue is lost during these periods, but its use as an energy substrate is limited. Amino acids that are produced from the breakdown of muscle are used primarily for the renewal of enzymes and other proteins essential to survival. Fat provides the primary energy substrate for human metabolism after a fast of two or three days. As fat mobilization increases, short carbohydrate molecules known as ketone bodies accumulate as byproducts. The brain uses a special set of enzymes to metabolize these molecules rather than relying on fat metabolism directly, since long-chain fatty acids do not cross the blood-brain barrier efficiently. The accumulation of ketone bodies, including acetone, in the blood can give the breath and urine of a starving person a distinctive airplane glue odor.

The ability to store and mobilize fat is particularly useful in an ecology where energy availability in the environment fluctuates above and below what is needed to survive. It allows an organism to use the excess energy available in times of plenty to survive times of scarcity. If there are never any times of scarcity, if energy is always available in excess of need, then there is no incentive to develop and deploy mechanisms for its storage and retrieval. We don't have mechanisms to store and retrieve oxygen, for example. Similarly, if energy is never available in excess but is always insufficient to meet all current needs, then storage is not a solution. Efficient allocation algorithms and the ability to cut metabolic corners are then more to the point. The fact that humans of both sexes have such a well-developed capacity for energy storage suggests that formative human environments were typically of the boom and bust variety. There is even some evidence that recent history may have intensified this aspect of the ecology of certain human populations, with the result of an increased selection pressure for fat storage capacity. Many Polynesian populations are presumed to have faced long periods of scarce energy availability (on long ocean crossings, perhaps) as well as periods of relative superabundance (early colonization phases of new islands). Many of these same populations have a dramatic energy storage capacity, so that obesity and diabetes become serious problems when commercial economies create conditions of chronically high energy availability. The Univeristy of Michigan geneticist James Neel first advanced this explanation, calling it the thrifty gene hypothesis.

Fat reserves also allow us to meet a fluctuating physiological demand, even when the demand temporarily exceeds environmental supply. Glycogen serves this function in the short term, allowing for muscular and biochemical activity long after the traces of our last meal have vanished. But where glycogen can only meet demands measured in minutes and hours, fat can meet demands measured in weeks and months. As we have seen, this ability to mobilize stored fat, our central mechanism for coping with times of food scarcity, is also central to a woman's ability to meet the metabolic demands of late pregnancy and intensive lactation. The physiological capability for efficient fat storage may have evolved originally in order to allow a relatively constant metabolic demand to be met in the face of fluctuating environmental supply. This capability is demonstrated by both sexes. Once developed, the capability may have been enhanced in females in order to allow a relatively constant supply to meet a variable metabolic demand. If it evolved first in the service of enhanced survival, fat storage appears to have gained a second function in the service of reproduction.

Frisch was the first person to incorporate this role of fat storage into her hypothesis regarding female menstrual function. In tracing the development of that hypothesis in the previous chapter, we saw that fat per se was not originally identified as a key variable, nor was adult menstrual function a part of the original hypothesis. Both of these elements of Frisch's hypothesis emerged from her efforts to meet specific criticisms. Since arriving at them, however, Frisch has defended her contention that a minimum level of fatness is necessary for regular menstruation to occur. Examples are cited associating people who have a particularly lean body composition with amenorrhea, including anorexics, athletes, and ballet dancers. The original Frisch and McArthur data showing the recovery of menstruation in previously anorexic women after they experienced weight gain has been complemented by anecdotal accounts of athletes and dancers who begin menstruation when their training regimes are broken. Potential physiological pathways by which fat storage may influence menstruation have also been proposed. In all its presentations and applications, however, the Frisch hypothesis focuses on one particular independent variable—stored fat—and one particular dependent variable—presence or absence of menstruation. In order to examine the relationship of energy to female reproduction properly, each of these variables needs to be placed in a more complete context.

ACCOUNTS AND BALANCES

Energy storage and utilization is a dynamic process. At any point in time a woman may be characterized by the amount of stored energy she has on her body, but that is not a complete description of her energetic condition. It is a measure of current energy status in the same way that a bank account statement is a measure of a person's current financial status. But two women who have exactly the same amount of stored fat can be in very different energetic conditions. One may be losing weight and fat, for example, while the other may be gaining. These are differences of energy balance rather than energy status, one woman being in negative energy balance and the other in positive energy balance. Quantitative differences in energy balance are possible as well as qualitative ones. Two women may have the same amount of fat and both may currently be losing weight, but one may be losing weight and fat faster than the other. The absolute value of the difference between energy expenditure and energy intake is greater in the woman who is losing weight faster, so that she is experiencing a greater degree of negative energy balance than the woman who is losing weight more slowly. In the financial analogy, energy balance corresponds to the net difference between income and expenses. Two people can have the same amount of money in the bank, but one person may be adding to her savings each month while the other is drawing down. As Mr. Micawber says in *David Copperfield,* "Annual income twenty pounds, annual expenditure nineteen nineteen six, result happiness. Annual income twenty pounds, annual expenditure twenty pounds ought and six, result misery." Micawber's distinction is a qualitative one between positive and negative financial balance. However, the degree of misery or happiness can presumably also be influenced by the magnitude of the imbalance between income and expenses.

Finally, two women can have the same amount of stored fat and the same direction and level of energy balance and yet differ in the rate at which energy is flowing through their bodies. This aspect of energetic condition, energy flux, is perhaps the most subtle to understand. Consider, though, the difference between a woman suffering through a famine in northern Africa and her kinswoman training on a different continent to compete in the Olympic games. Both may have comparable low levels of stored fat and both may be currently in neutral energy balance, neither gaining nor losing weight or fat. The famine victim, however, is in low energy flux: she has very low energy intake, but has reduced her energy expenditure to match, both by reducing her physical activity and by reducing her basal metabolism. The athlete is in high energy flux: her energy expenditure is very high owing to her training regime, but her energy intake is high as well. In financial terms, the distinction is analogous to the difference between someone with a high monthly income and high expenses to match and someone with low income and expenses. Monetary flux would represent the rate at which funds flow into and out of their bank accounts each month. This flux can be distinguished both from the net difference between income and expenses (monetary balance) and the amount in the account at the end of the month (monetary status).

It is likely that natural selection has shaped a woman's reproductive physiology to be sensitive to more than one of these aspects of energetic condition. To continue the financial analogy, suppose that the question of allocation of energy to reproduction is analogous to the allocation of financial resources to building a new house. Having a large bank account might provide the necessary funds. Few of us ever have that much money in the bank at once, however. Failing that, having an income greater than current expenses may allow for funds to be allocated to the new construction expense category. In contrast, having current expenses that already exceed income may effectively reduce the ability of the bank account balance to support a new construction project. The reserve may be required just to meet other monthly expenses. The more steeply the account balance is being drawn down each month, the less likely it is to be able to cover the cost of the additional construction project. The more steeply it is increasing each month, the more likely it is to exceed current needs and provide for additional categories of expense in the future. Finally, financial flux may constrain the ability to meet additional construction costs as well. If I am already working two jobs to pay my current bills (mortgages, car payments, college tuitions), I may be hard pressed to generate an additional positive financial balance with which to build a house. By the same token, if I have already reduced my expenses to the bare minimum (sold my car, moved to a one-room apartment) to balance an externally imposed low income, my chances of freeing up sufficient funds to build a house may also be slim. Thus both high and low monetary flux may make financing additional building projects difficult. A moderate level of flux provides the greatest flexibility since it may be possible through extra effort to raise my income a bit, or through extra belt-tightening to reduce my expenditures.

Physiologically assessing the wisdom of embarking on a new pregnancy may be analogous to financially assessing the wisdom of starting a new major construction project. The level of current reserves, the balance of intake over expenditure, and the current level of resource flux may all contribute significantly to the probability of success. They may also be confounded with one another in ways that can be misleading. Extremes of energy status in particular are likely to occur in individuals with a recent history of either negative or positive energy balance, or with extremes of energy flux.

HOW LOW CAN YOU GO?

Menstrual status has similar limitations as a dependent variable in any attempt to understand the relationship between energetic condition and female reproduction. The importance of menstruation in the Frisch hypothesis is traceable to the original focus of that hypothesis on adolescent maturation. The onset of menstruation is an outward marker of reproductive maturation in girls that can be noted easily in longitudinal growth studies. Once the focus of the Frisch hypothesis shifted to adult reproductive function, menstrual status proved equally practical. Information on menstruation can be collected by survey and recall based on self-reports. Use of this variable, however, conveys an impression of female fecundity as an "on or off" variable. Yet although the absence of menstruation is almost certain evidence of negligible fecundity, the presence of menstruation is not evidence of full fecundity. For example, in the longitudinal studies of the resumption of ovarian function during lactation reviewed earlier resumption of menstruation was not synonymous with resumption of ovulation. And even with evidence of the resumption of ovulation in nursing mothers, differences persisted in the level and duration of luteal progesterone production.

Far from being a simple on-off phenomenon, ovarian function in women represents a graded continuum. It is "graded" in that discrete, qualitative states can be recognized that have functional significance; the presence of menstruation represents growth of the endometrial lining of the uterus, for instance, and the presence of a luteal phase rise in progesterone indicates ovulation. It is a "continuum" in that a quantitative scale of ovarian function can be established on the basis of ovarian steroid levels. The version of "the human menstrual cycle" that finds its way into most textbooks is usually drawn from the very top of this continuous scale. In this ideal menstrual cycle circulating estradiol increases exponentially during the first two weeks (the follicular phase of the cycle), culminating in a peak and then an abrupt drop leading to the LH surge that triggers ovulation. A steep rise in progesterone levels begins right after the LH surge and reaches a high plateau that is maintained for a week or so, followed by a steep decline in progesterone to basal levels over a period of days, ending in menstruation (the luteal phase of the cycle).

This textbook version of the menstrual cycle is not always encountered in nature, however. Variation from that ideal can be represented on a continuous scale of production of the two major ovarian steroids, estradiol and progesterone. If we work down from the textbook ideal, lower estradiol levels come first, a condition that can be referred to as follicular suppression relative to the textbook menstrual cycle. Follicular estradiol production reflects the growth and maturation of the follicle containing the developing oocyte. The bigger the follicle, the more estradiol it produces. Slowing the growth of the follicle and/or limiting its final preovulatory size can lead to lower circulating estradiol levels and possibly a longer follicular phase to the cycle before ovulation. Follicular growth may still be sufficient for ovulation and the formation of a corpus luteum, however.

Luteal suppression, reflected in lower progesterone levels after ovulation and/or a shorter period of progesterone secretion, comes next on the continuum. Because the corpus luteum is formed from the follicle that developed in the first half of the cycle, at some point follicular suppression is thought to blend into luteal suppression. The growth of the follicle and its estradiol producing capacity is driven in part by the proliferation of the granulosa cells it contains. If the proliferation of the granulosa cell mass is suppressed, the cellular machinery available for progesterone production in the second half of the cycle can also become limited. A reduced circulating level of progesterone is a sign of this. It is interesting to note, though, that progesterone is produced both by luteinized versions of the granulosa cells of the follicle and by luteinized versions of the theca cells that surrounded the follicle before ovulation.

These two cell types become intermingled during the transformation of the follicle into a corpus luteum. But they appear to retain important differences in function. In particular, luteinized theca cells retain the responsiveness to LH, which they displayed before ovulation, while luteinized granulosa cells are much less responsive. The luteinized theca cells produce progesterone in episodic bursts or pulses, possibly related to LH pulses from the pituitary, while the luteinized granulosa cells produce progesterone in a continuous, sustained fashion known as tonic production. Luteinized theca cells are also, not surprisingly, the cells that respond to hCG produced by an implanting embryo in the process of corpus luteum rescue. (Remember that hCG and LH are very similar in structure and operate through interaction with the same receptors.) Luteinized theca cells appear therefore to have a key role in responding to a pregnancy if it occurs. Quantitative differences in progesterone levels, however, appear to be due principally to differences in tonic progesterone production by the luteinized granulosa cells, as we shall see. This is the aspect of luteal progesterone production that would be most affected by any suppression of the proliferation of the granulosa cell mass of the follicle.

Variation in luteal phase length is much less than variation in follicular phase length. Follicular phase length appears to be determined by the length of time it takes a developing follicle to reach its maximum size. The slower it grows, the longer it takes. Luteal phase length is much less flexible. Empirically, however, luteal phases with lower progesterone levels tend to be somewhat shorter in their duration than those with higher levels. This means that at some point, longer follicular phases become associated with shorter luteal phases. The net effect of these opposite changes on the overall length of the menstrual cycle may be negligible. Hence regular menstruation may persist even at this level of ovarian suppression.

As suppression of ovarian function becomes even more profound, ovulation may fail. The lack of a significant progesterone rise ordinarily provides evidence of this failure when it occurs. Anovulatory cycles may still end in menstruation, however. Follicles may have begun the course of maturation without completing it, providing enough estrogen stimulation to cause proliferation of the uterine lining. In the absence of progesterone support, such an endometrium is eventually sloughed off in menstruation. Exact coincidence of such "anovulatory menstruation" with the expected date of menstruation is unlikely. It might happen sooner or later than is typical for an ovulatory cycle in a given woman. Hence irregular menstruation may often token ovulatory failure, particularly if it becomes frequent. Occasional anovulatory cycles may go unnoticed, however.

If follicular development is suppressed even further, estradiol production may become insufficient to stimulate endometrial proliferation, or the proliferation may be so limited that the uterine lining is eventually resorbed rather than shed. This may occur sporadically, with follicular development initiated unsuccessfully one or more times before proceeding far enough to cause menstruation or to achieve ovulation. In such cases the interval between successive menses almost certainly lengthens. If the interval approaches two months or more between menses, the condition is referred to as oligomenorrhea and the interval itself is said to contain one or more "missed" periods.

The final level of ovarian suppression, amenorrhea, involves complete cessation of follicular development and associated estrogen production. Menses cease altogether.

The presence or absence of menstruation, without regard to its regularity, discriminates only among the most profound levels of ovarian suppression on this continuum. The highest levels on the continuum, by contrast, cannot be reliably discriminated on the basis of menstrual patterns at all. Follicular suppression, luteal suppression, and sporadic ovulatory failure all require hormonal measurements for their determination. These subtle degrees of ovarian suppression have been associated with decreases in fecundity, however. In a longitudinal study of twenty-four healthy Boston women who were trying to become pregnant, small variations in follicular estradiol levels were found to significantly affect the probability of conception. This was true both between women (women with higher follicular estradiol levels were more likely to conceive in any given month) and within women (an individual woman was more likely to conceive in a month when her follicular estradiol levels were high for her than in a month when they were low). The variation in estradiol levels associated with a significant increase in the probability of conception was relatively modest and not associated with variation in luteal progesterone levels.

It is more difficult to be sure that variation in luteal progesterone is associated with differences in fecundity. Some studies of natural conception have found higher progesterone levels early in the luteal phases of conception cycles than in nonconception cycles. But by the time one can reliably detect differences in progesterone production, pregnancy may already be affecting those same levels via the production of hCG by the embryo. There are additional reasons, however, to believe that luteal suppression lowers the probability of conception. Women who have chronically low progesterone levels, or chronically short luteal phases, often have trouble conceiving or sustaining a pregnancy, a condition known as luteal phase deficiency, or LPD. Progesterone supplementation can increase conception rates considerably in many of these women. Progesterone supplementation also increases rates of conception in older women who have low progesterone as a consequence of age.

Irregular menstruation has also been associated with a lower probability of conception than regular menstruation both between and within women. A longitudinal study of 295 women attempting to conceive found that the greater the degree of cycle length variability a woman experienced, the less likely she was to get pregnant, and the greater the departure of an individual cycle's length from a woman's overall mean length, the lower the probability of conception in that cycle. In this study urinary hCG levels were also measured to detect early pregnancy loss. Cycle length variability did not affect the rate at which pregnancies were lost, but affected only the rate at which they were initiated. As noted above, irregular cycle length (as opposed to consistently long or consistently short cycles) is often a sign of irregular ovulation, which presumably contributes to the lower level of fecundity.

When menstruation becomes irregular enough to be classified as oligomenorrhea, the association with low ovulatory frequency is taken for granted.

Thus the continuum of ovarian function is also a continuum of fecundity, with even subtle suppression of ovarian steroid levels associated with detectable decreases in the probability of conception. Individual women may move back and forth along this continuum at different points in their lives. Far from being a simple on-off variable, ovarian function is a continuous variable reflecting higher or lower probabilities of conception. Energetic condition is not simply a matter of fat reserves either. Equipped with a more sophisticated understanding of both female energetics and ovarian function, we are now in a better position to consider how energetic conditions may affect female fecundity.

SLIM GYM

Studies of women who exercise regularly have provided one window on the relationship of energetics and ovarian function. The first evidence of a link between exercise and suppressed ovarian function came from observations of high frequencies of amenorrhea and oligomenorrhea among female distance runners on college track teams. The frequency of amenorrhea (defined as three or fewer periods of menstrual flow in twelve months) varied with weekly training mileage, from 6 percent among those running less than ten miles per week to more than 40 percent among those running eighty miles a week or more. Numerous other studies have confirmed the increased prevalence of irregular menstrual patterns and the associated suppression of reproductive hormones in women engaged in regular aerobic exercise of various forms. Although athletes as a group are relatively lean, the incidence of amenorrhea in female athletes has not been found to correlate with body fatness. This suggests that it may be the high energy flux experienced by athletes, rather than low energy status, that is associated with ovarian suppression.

Other studies have probed the issue further and provided evidence of a suppressive effect of exercise on ovarian function even in women who continue to menstruate regularly. One of the first of these studies reported lower luteal progesterone values in a single woman runner in months when she was engaged in regular training than when she was not training. The length of the luteal phase of her cycle also decreased progressively as weekly training mileage increased, even though overall cycle length did not change. So rather than there being an abrupt change in ovarian function near some threshold level of exercise stress, moderate amounts of running appeared to be associated with mild degrees of ovarian suppression, with increasing degrees of suppression as exercise levels increased.

There is a possibility that the observed tendency to ovarian suppression among female athletes is a result of a subject selection bias. Competitive athletes are not, after all, a random sample of the population. Their success at their chosen sports might reflect constitutional factors that facilitate athletic excellence but that, at the same time, are disruptive to regular ovarian function. High androgen levels, for example, could contribute to linear body build, lean body composition, and high aerobic capacity while at the same time inhibiting normal ovarian function. In support of this notion, it has been noted that the mothers of female athletes report later menarcheal ages than the average for their cohort, suggesting a heritable component to the reproductive physiology of athletic women. Thus the fact that athletic activity and amenorrhea are associated doesn't necessarily mean that the one causes the other.

Beverly Bullen together with Janet McArthur and other colleagues designed a study to test the hypothesized causal relationship. They sought to determine whether experimentally assigned exercise would induce menstrual irregularity and disrupted ovarian function in previously inactive women. At the same time they tried to separate the effects of high energy expenditure (energy flux) from negative energy balance and low energy status. In other words, if exercise is associated with ovarian suppression, is it the exercise itself that is the cause, or does ovarian suppression occur only if the exercise leads to weight loss or low fat reserves?

A group of twenty-eight women was brought to a summer camp in

New England and subjected to a regime of regular exercise over a period of two months. Exercise included running, starting at a level of four miles a day and increasing to ten miles a day by the fifth week, as well as two hours a day of other aerobic activity (such as tennis, swimming, and volleyball). Food intake was adjusted to maintain body weight in twelve of the women despite the high levels of energy expenditure, and to allow weight loss at the rate of 0.45 kilograms a week in the remaining sixteen women. The frequency of menstrual irregularity, ovulatory failure, and suppression of luteal progesterone increased dramatically in both groups in the first month. The frequency of the more severe forms of ovarian disruption increased further in the second month among women in the weight loss group but not among those in the weight maintenance group. After the two-month study the activity levels of all subjects decreased and normal menses resumed within three months. The results of this study imply that exercise can indeed cause ovarian suppression, even when energy balance and status are maintained. The level of activity used to obtain these results, however, was quite high and the change from the subjects' previous activity levels quite abrupt. Thus the study leaves open the question of whether lower levels of exercise would have milder suppressive effects, and whether some accommodation to a new level of physical activity would occur over time.

Studies of the relationship between recreational running and ovarian function help to answer these questions. In one study of eight women between twenty-three and thirty-five years of age who ran an average of 12.5 miles a week but were of stable weight, normal weight for height, and experiencing regular menstruation, progesterone profiles were significantly lower than among nine nonexercising controls matched in age, height, and weight. A second study of twenty-two women (also twenty-three to thirty-five, with stable, normal weight, and regularly menstruating) who ran an average of 25 miles a week confirmed this result. Salivary estradiol levels were also suppressed in these women relative to the levels in controls. In this case the subjects had been engaged in the same level of activity for at least three years.

Taken together, the results of these studies indicate that the high energy flux of exercise can cause suppression of ovarian function even when it is isolated from negative energy balance or low energy status. It also appears that exercise affects ovarian function in a dose-response manner: the more intense the exercise the greater the suppression. At very high levels of energy flux ovarian suppression may occur to the point of disrupting menstrual regularity. But even at moderate levels, exercise can cause measurable suppression along the continuum of ovarian function. It also appears that there is no long-term accommodation of ovarian function to levels of exercise. Ovarian suppression seems to persist as long as a chronic level of exercise is maintained and to be readily reversible when the exercise regime is discontinued. More studies of long-term exercise effects are needed, however, to be sure of this. Finally, we should note that the suppressive effect of exercise on ovarian function is not limited to young women. Many of the studies of exercise and ovarian function have focused on college-age athletes since they represent a readily recruitable pool of potential subjects. But studies of older women in the prime of their reproductive years show similar effects.

Fewer studies have attempted to isolate negative energy balance as a potential cause of suppressed ovarian function. In the study of Bullen and her colleagues described above, it is notable that ovarian suppression was more profound and more progressive in women who lost weight while exercising than in those who maintained their weights. In a rather dramatic study, M. M. Fichter and K. M. Pirke followed five young women during a three-week total fast. Three of the subjects regressed to a prepubertal LH pattern before the end of the fast, and none of the subjects menstruated, either during the three-week fast or during the six week follow-up period. Subsequently, Pirke and colleagues studied the hormonal profiles of young women on calorie-restricted diets (1,000 kilocalories a day) over the course of six weeks. Secretory patterns of LH were normal, but the incidence of amenorrhea and luteal phase defects (short and low luteal progesterone profiles) increased during the study. In a follow-up study of calorie-restricted vegetarian diets the degree of luteal suppression was observed to increase with greater weight loss, and to be greater in younger women (nineteen to twenty-four years old) than in older women (twenty-five to thirty years old).

A different study followed eight women who were voluntarily restricting their caloric intake to lose weight and nine stable-weight controls. All the subjects were within the normal range of weight for height throughout the study period. On average, they lost two kilograms per month, a relatively moderate rate. Yet their progesterone levels during months of weight loss were significantly lower than during periods of stable weight, and the degree of luteal suppression was correlated with the amount of weight lost. The effects of weight loss appeared to persist into the following month as well, so that the progesterone levels in the month after weight loss were even lower than those during the month of weight loss itself.

Even moderate, unintentional changes in energy balance appear to be associated with variation in ovarian function. In the study of women attempting to become pregnant referred to above, a significant correlation was evident between variation in a woman's weight and variation in her follicular estrogen levels. Slight changes in weight of 2 percent or less in either direction were associated with significant parallel changes in estradiol levels. These same changes in follicular estradiol were found to significantly affect the chance of conceiving in a given month.

It appears, then, that negative energy balance is associated with ovarian suppression even when isolated from low energy status or high energy flux. It also appears that the effect is dose-dependent, with greater degrees of negative energy balance associated with greater degrees of ovarian suppression. The study by Bullen and her colleagues suggests that the effects of high energy flux and negative energy balance can be compounded when the two conditions coincide, but few studies have focused on these sorts of interaction effects.

THE FRUITS OF OUR LABOR

As illuminating as the studies of dieters and athletes are regarding the interaction of energetic conditions and ovarian function, it is difficult to extrapolate from these rather specialized situations to the realm of general human ecology. Arguments have been advanced suggesting, for example, that the low fertility of !Kung hunter-gatherers may be associated with the heavy energy expenditure assumed to be a consequence of their foraging activities and nomadic lifestyle, citing the evidence of reproductive suppression among athletes by way of evidence. It has also been suggested that traditional divisions of labor between the sexes may derive from the negative impact of demanding aerobic activity on female fecundity. It is difficult, however, to find true analogs among privileged, Western populations for the energetic conditions that are likely to have been formative during human evolution, or indeed that exist for most natural fertility populations today. In order to investigate the responsiveness of ovarian function to more representative energetic stresses it is necessary to move studies of reproductive physiology out of the comfortable surroundings of the laboratory and the clinic and into the environments where people live their lives. In these contexts, energy intake and expenditure, balance and flux, are often not consequences of individual choice or conscious manipulation. They are more often the unavoidable consequences of trying to make a living and provide for one's family. Studying the effects of these ecologically imposed realities of human energetics allows us to think about their reproductive consequences with more confidence. At the same time we usually lose the level of experimental control that may be possible under more artificial circumstances. The trade-off is worth it, however, for the deeper insight it yields. Several studies that my colleagues and I have conducted around the world help to demonstrate this broader approach.

The Lese of the Congo's Ituri Forest are subject to chronic as well as acute energetic stress as a consequence of their subsistence ecology. The bulk of their caloric intake derives from the produce of their gardens, primarily cassava, maize, plantains, rice, and peanuts. Although there may be temporary abundance following a harvest, averaged over the year caloric intakes are low. The relatively short stature of the Lese (mean for men = 162 centimeters, for women = 153 centimeters) reflects moderate, chronic undernutrition. The horticultural cycle of the Lese is strongly controlled by the short dry season in December and Jan-

Lese women preparing food.

uary, during which new gardens must be cleared, burned, and planted, resulting in harvest seasons for different crops between June and November. Depending on the size of the harvest, these resources are progressively depleted through the winter and spring, leading to a typical pattern of negative energy balance between December and June. As is often true of ecological research, however, typical patterns are only borne out in the long term and significant year-to-year variations exist. In 1984, for example, the summer harvests were so meager that most individuals continued to lose weight well into the fall, and the average weights in December were even lower than those in the preceding June.

In addition to the energetic stresses that derive from their ecology, the Lese are subject to high disease burdens, including directly transmissible diseases such as measles, water-borne diseases such as giardiasis, vector-borne diseases such as filariasis and malaria, and sexually transmitted diseases such as chlamydia and gonorrhea. The energetic costs of chronic disease and parasite burdens are difficult to estimate, but are certainly not trivial. In addition, some diseases, such as malaria and gonorrhea, also have a direct impact on reproductive physiology, compromising placental oxygen transport in the case of malaria and causing scarring and blockage of the fallopian tubes in the case of gonorrhea. The toll of venereal disease on the fertility of older cohorts of Lese women can be seen in the high frequencies with which women report a last birth at an unusually young age, a situation that is common among other populations in sub-Saharan Africa as well. Younger cohorts of Lese women show less evidence of primary or secondary sterility, however. The reproductive histories of these women are marked instead by long interbirth intervals despite a tendency to wean children at about one year of age. As a consequence of all these factors, women spend a relatively large fraction of their reproductive lives in a state of cyclic ovarian function, rather than in gestation or lactational amenorrhea, despite the absence of effective artificial birth control. These observations suggest that the average waiting time to conception may be longer among the Lese than is typical among Western women.

Associated with the evidence of low fertility among the Lese are data indicating low levels of ovarian function relative to Western standards. Average profiles of salivary progesterone in samples collected from regularly cycling women in three different years, 1983, 1984, and 1989, are all significantly lower than the average profile for Boston women. So are ovulatory frequency and estradiol levels in all phases of the menstrual cycle. The menstrual cycle tends to be longer among the Lese than among Boston women, though not significantly so. The duration of menstrual bleeding, however, is significantly shorter, perhaps reflecting a lesser degree of endometrial proliferation as a result of lower estrogen stimulation.

The generally low levels of ovarian function observed among the Lese may contribute to their low fecundity. But we cannot conclude on the basis of population comparisons alone that their low ovarian function is a consequence of energetic stress. It might reflect specific disease consequences, or result from specific nutrient deficiencies, or even represent genetic differences between broadly separated populations. We can, however, examine patterns of variation in ovarian function within the Lese population, between different women and within women over time, to consider the extent to which they are associated with variation in energetic stress.

There is scant evidence that variation in energy status alone, separated from energy balance and energy flux, is associated with variation in ovarian function among the Lese. Progesterone levels, for example, do not show a correlation with weight or body mass index (BMI, a measure of relative weight corrected for height) among women measured in the same season. Nor is there any indication that estradiol levels or patterns of menstrual bleeding are associated with between-woman differences in weight or BMI.

Lese ovarian function does, however, show evidence of significant variation associated with energy balance. In 1984, for example, poor harvests resulted in a continuing decline in average weight among Lese women into October. As the average weight of the population declined, so did average indices of ovarian function, such as ovulatory frequency. Within the population, women who lost more than two kilograms over this period had lower progesterone levels and a lower frequency of ovulation in October than did those who lost less or gained weight.

In 1989 Gillian Bentley and Alisa Harrigan followed Lese women through the period of diminishing food supplies preceding the peanut harvest in June. Average weights progressively declined from February to June with evidence of a turnaround in July. Both hormonal and menstrual indicators of ovarian function showed parallel patterns of variation: cycle lengths increased, durations of menstrual bleeding decreased, and ovulatory frequency declined, with the trends in cycle length and ovulatory frequency reversing after the harvest.

Among the Lese, then, evidence suggests that negative energy balance is associated with suppression of ovarian function. In this case, however, negative energy balance is not the result of self-imposed exercise or dieting regimes, but a result of local subsistence ecology. The effect is normally seen during the "hunger seasons" preceding the June harvest, but can occur at other times of the year as well if food supplies dwindle.

Another, very different population has also provided data on energetic condition and ovarian function in a natural setting: the Tamang of central Nepal. The Tamang subsist on a combination of agriculture and pastoralism in the foothills of the Himalayas at altitudes ranging from 1,350 to 3,800 meters. Rainfall patterns are strongly seasonal with three quarters of the annual average of four meters of rain falling during one quarter of the year, the three-month monsoon season from mid-June to mid-September. The Tamang grow crops of wet rice, millet, wheat, and barley on terraced fields at successive altitudes and herd mixed flocks of sheep, goats, oxen, and cattle between high and low pastures. Tamang women work hard at all seasons of the year, particularly during the monsoon, when long hours must be spent in cooperative labor groups transplanting rice. Unlike the Lese, however, the Tamang have food available quite constantly throughout the year. Variations in workload, rather than variations in caloric intake, then, are more likely to produce individual variation in energetic condition.

Catherine Panter-Brick led a study of patterns of ovarian function in a group of Tamang women during the late fall of 1990 (October to November, when workloads are relatively low), and again during the following monsoon (August to September, when workloads are relatively heavy). Saliva samples were collected every other day and analyzed for progesterone and estradiol concentrations in twenty-one nonpregnant, nonlactating women who were menstruating regularly.

Energy balance among the Tamang is much more variable than among the Lese, perhaps because it is more a consequence of factors that may vary greatly between women, such as energy expenditure in work, than it is a consequence of common food shortages. Weight changes between the two seasons for individual women ranged from a loss of 2.8 kilograms to a gain of 4.8 kilograms, but the average weight for the entire sample did not change significantly.

The progesterone data indicate that ovarian function is lower, on average, during the monsoon than during the late fall. Average progesterone values over the whole luteal phase decline during the monsoon season by 27 percent, and average values in the mid-luteal phase de-

Tamang women in their fields.

cline by 38 percent. The proportion of cycles judged ovulatory for the whole sample also declines from 71 percent in the late fall to 38 percent during the monsoon. Dividing the sample into those with a net gain in weight between the seasons and those with a net loss makes it clear, however, that the seasonal changes in progesterone levels are associated with negative energy balance. Among the eleven women who suffered a net weight loss between the two seasons average luteal progesterone levels declined by 40 percent, and mid-luteal levels by 50 percent, while the same levels remained unchanged in the ten women who showed a net weight gain. Like the patterns of energy balance among the Lese, those among the Tamang derive from their subsistence ecology, in this case variations in energy expenditure more than in energy intake, but with the same demonstrable association of negative energy balance with suppressed ovarian function.

A final example of variation in ovarian function associated with energetics can be drawn from yet another continent. Grazyna Jasienska studied salivary progesterone profiles among a group of farm women in rural, southern Poland. In the mountain valleys outside Cracow agriculture is still largely unmechanized. Families plant cereal crops and potatoes and graze dairy cattle on fields and pastures scattered over many kilometers. All able-bodied family members, including adult women, contribute their labor on a daily basis to subsistence activities. Workloads are considerable throughout the year but also vary substantially according to season. The summer haying season, for example, requires long hours of travel by foot and manual labor, turning the hay daily with rakes so that it can dry, gathering the hay by hand and loading it onto wagons pulled by draft animals, and transferring it into the barn for winter silage. During the winter, travel is restricted and work is more confined, although still lasting from before dawn until after dark. Food is abundant, however, at all seasons, and the nutritional status of the women is well above that of the women in the Congo or Nepal, comparable to that of women in Boston. Their high workloads place them in a situation of relatively high and variable energy flux. Yet, as with the Lese and the Tamang, the energetic stress faced by rural Polish farm women is a consequence of their subsistence ecology, not a matter of whim.

In general, ovarian function among these Polish women, as indexed by salivary progesterone profiles, shows the impact of workload variation both within and between women. In an initial study women were placed into one of two groups according to their observed summer workload. Those in the higher workload category had significantly lower levels of progesterone than did those in the moderate workload category. In a subsequent study, women were followed through the summer into the fall. Average progesterone levels were low during the high workload months of July and August and rose during the lighter workload months of September and October. Estimated energy expenditure in July and August was negatively correlated with progesterone levels in those months, but positively correlated with the increase in progesterone levels between July and October. Thus the harder women work during the summer months the greater the suppression of progesterone compared to their own levels in the fall, when work is light. Measures of food intake and energy status (in this case percentage of body fat) were taken on the women during the same periods. When workload increased, so did energy intake. Changes in energy status, on the other hand, were negligible and showed no independent correlation with ovarian function.

The results of the Polish study are particularly important for a number of reasons. They show that energy flux, independent of energy balance or energy status, can influence ovarian function, even when it is associated with subsistence work rather than high-intensity aerobic exercise. In addition, they show that suppression of ovarian function along its continuum can occur in association with energetic stress even in well-nourished women. Such suppression does not occur only in women living under marginal energetic circumstances.

Together these field studies demonstrate that human ovarian function varies with energetic stress in similar and predictable ways across a broad range of ecological, geographical, and cultural settings. The patterns of variation in ovarian function we observe in response to the energetic constraints and consequences of local subsistence ecology in places like the Congo, Nepal, and Poland are similar to the patterns we observe in Boston women under more idiosyncratic circumstances. The consistency of the relationship between energetic condition and ovarian function suggests that it is a general feature of the reproductive biology of our species. The fact that ovarian function varies in a quantitative, dose-response manner across a range of levels of energetic stress that are typically encountered by human beings in the course of their daily lives suggests that it represents functional modulation and not pathological failure of homeostasis in the face of extreme or unusual stress.

EVERYTHING IN ITS SEASON

It is a more difficult task to determine whether the observed variation in ovarian function associated with subsistence ecology has implications for female fecundity. Biological fecundity is extremely difficult to measure while the achieved fertility of individuals is profoundly influenced by a number of other factors, many of them much stronger and more direct than energetic constraints on ovarian function. Comparison of fecundity between groups or populations requires aggregate data on birth interval components from large samples of individuals, data that are rarely available for populations of interest, and even then the estimates of fecundity are questionable. Variation in fecundity within populations, however, and its potential relationship to general, population-wide variation in energetic conditions, can be investigated. Often one of the clearest manifestations of such variability is seasonality in the pattern of human births.

Seasonal variations in the pattern of human births are widely observed, indeed virtually universal. Any number of factors, either directly or indirectly related to seasonal variations in climate, have been suggested as possible causes of human birth seasonality. Some researchers have suggested direct effects of variations in light or temperature on male or female reproductive physiology, but these hypotheses have generally not held up very well. More often it is assumed that seasonality of births reflects seasonality of behavior with peaks in coital frequency that are culturally entrained. Many Anglophone countries, including the United States, Canada, and New Zealand show birth peaks in September, for instance, nine months after the Christmas–New Year's holidays, while many European populations, including Britain and France, show birth peaks in April, nine months after the traditional August vacations. In non-Western populations as diverse as arctic Eskimos and New Guinea highlanders, cultural cycles linked to the annual round of subsistence activities have been cited as the probable causes of birth seasonality. In other populations, seasonal separation of spouses has been related to seasonality in conceptions and births.

It is very likely that birth seasonality arises in different populations for different reasons. There is no need to assume that a single explanation will serve for all instances. Our consideration of energetic constraints on human ovarian function suggests, however, that where natural fertility populations are subject to strong seasonal variations in energy balance we should expect to find an effect on the seasonality of conceptions and births. The evidence for such an effect is in fact widespread, particularly in populations relying on subsistence agriculture, though it is not often recognized. It is instead usually assumed that seasonal patterns of intercourse are responsible for birth seasonality in these populations, even in the absence of any supporting data.

A particularly illustrative example of birth seasonality in an agricultural population, with a particularly careful analysis, is provided by longitudinal studies carried out in the vicinity of Matlab, Bangladesh, under the aegis of the International Center for Diarrhoeal Disease Research. The Matlab region lies forty miles south of Dhaka within the Ganges-Meghna floodplain. Climatological variation is dominated by the seasonal pattern of the monsoon rains, which are heaviest in the months of July to September, leading to seasonal flooding of the low-lying agricultural fields. The staple crop is wet rice, harvested at the end of the monsoon in November. Seasonal variation in food availability together with seasonal variation in workload produces strong seasonal variations in nutritional status among the women, weights being the lowest during the monsoon, before the rice harvest.

Birth seasonality was first reported for this population by J. Stoeckel and A. Chowdhury with a peak in November and December. An initial prospective study of birth interval components indicated that the monthly probability of conception varied seasonally with a peak in April and a nadir in July. The same study also indicated that lactating women were most likely to resume menstruating in November irrespective of the month in which they had given birth, and that women who resumed menstruating between November and April had shorter waits to their next conception than women who resumed menstruating in the opposite half of the year.

In a study designed to probe the causes of birth seasonality further, Stan Becker and his colleagues completed an analysis of birth seasonality in the Matlab area based on a larger sample size of approximately 2,500 women followed prospectively for four years. Information was collected monthly on recognized pregnancies and pregnancy outcomes, the introduction of liquid and solid supplements to the diets of breastfed children, times of weaning, resumption of menstruation postpartum, and the interval since last intercourse. The data were analyzed to address four specific questions.

1. Is the seasonal pattern of conceptions a result of seasonal resumption of menstruation?
2. Is it a result of seasonal patterns of intercourse?
3. Is it a result of seasonal patterns of fetal loss?
4. Is the seasonal resumption of menses a result of a seasonal pattern of weaning?

Each of the four questions in essence represents a hypothesis regarding the possible mechanisms underlying the observed birth seasonality. In the subsequent analysis, all four hypotheses are rejected. A peak in conception rates is observed between January and April for all women, regardless of the length of time they had been menstruating. Frequency of intercourse shows little seasonality and no correlation with fecundability. Fetal loss and stillbirth rates show little seasonal variation. Nor is the peak in resumption of menses postpartum clearly related to a peak in weaning or changes in infant diets. The authors are forced to conclude that the observed birth seasonality must not be a consequence of seasonal changes in sexual behavior or nursing behavior, or an artifact of seasonal patterns of resumption of menses postpartum or seasonality in fetal loss. Rather, they conclude that some mechanism that occurs between the time of insemination and the recognition of pregnancy must be susceptible to seasonal variation, perhaps the quality of the sperm, the quality of the embryo, or the quality of the ovum and maternal environment for fertilization and implantation. In particular they note that variation in maternal nutritional status might play a role. Resumption of menses is most likely immediately after the rice harvest, when energy balance shifts from negative to positive, and conception rates rise while women are gaining weight.

The Turkana of northern Kenya provide another example of birth seasonality in a different ecological context. The Turkana are traditionally nomadic pastoralists, herding cattle, goats, sheep, camels, and donkeys on the sub-Saharan savanna. Seasonality of rainfall once again dominates as a source of climatological variation. Nearly half the annual rainfall occurs during the months of March, April, and May, with significant rain beginning in February and continuing in June and July. Savanna grasses grow vigorously after the start of the rains in February. The rest of the year in contrast is very dry, and suitable grazing for the herds becomes progressively more scarce. Many Turkana have adopted an agricultural lifestyle instead, living in settlements southwest of Lake Turkana. Those who continue to pursue a pastoral lifestyle, however, show a pronounced seasonality in births, with over half the births occurring between March and June.

Paul Leslie and Peggy Fry studied the birth patterns recorded for the Turkana over a number of years and note that monthly variation in conceptions is highly correlated with monthly variation in rainfall three and a half months earlier. They suggest several mechanisms that could link the onset of the rains to the peak in conceptions. The cows calve at the beginning of the rainy season and milk, a staple of the Turkana diet, rebounds from its lowest availability to reach a peak at the height of the rainy season. Increasing milk availability could affect the frequency of conceptions either by increasing female nutritional status or by providing alternative food for nursing infants and thus lowering the metabolic load of lactation on the mothers. More plentiful grazing could lead to lower workloads and less frequent spousal separations. Other mechanisms are considered as well, but data to test the various possibilities are wanting. Without data on the intervening links in the putative causal chain linking rainfall patterns to births, the correlation is difficult to interpret. James Wood has pointed out that any two seasonal variables, storks and babies for example, will always be correlated with each other given an appropriate lag time, and that the correlation itself needn't imply any causal connection, direct or indirect.

In both the Bangladesh and Turkana studies it is at least possible that a well-documented pattern of birth seasonality is related to seasonal variations in female nutritional status. In Bangladesh, alternative mechanisms have been explicitly considered and rejected. But in neither case has the hypothesis of energetic seasonality driving birth seasonality been developed and tested in any detail. For the Lese and Efe of the Congo, on the other hand, a more specific model of birth seasonality has been developed, linking patterns of rainfall to patterns of human births through intervening links of garden size and productivity, human nutritional status and energy balance, and ovarian function.

In the Ituri Forest, as in many of the world's rain forests, rainfall is not constant through the year but is in fact quite variable. It is variable both between seasons and from year to year. The timing and duration of the all-important dry season are, moreover, not fully predictable, and significant year-to-year variations occur. These variations can have demonstrable negative impact on the size and productivity of Lese gardens. If the dry season ends prematurely before burning of the new slash has been completed, smaller gardens may result, or, in the worst case, it may be necessary to replant the previous year's garden plots despite significant depletion of soil fertility. If the dry season extends too long, seed and root germination in the new gardens may be poor. Either scenario can lead to significant variation in garden sizes and yields between years.

The limited yields from this system of slash-and-burn horticulture together with the highly synchronized harvests and the poor conditions for food storage in the tropical rain forest combine to produce strong seasonal variation in food availability. Garden produce from new harvests begins to become available in June and generally remains plentiful until the dry season. After the dry season, however, food supplies dwindle and are not replenished by new harvests. The Lese population comes to subsist more and more heavily on cassava, a nutritionally poor root crop, until the new peanut harvest the following June. Wild game and other forest resources are also relatively scarce during the period of low food availability. As we have seen, the preharvest "hunger season" is often accompanied by widespread weight loss.

Year-to-year variation in the severity of the hunger season occurs as well, occasioned by annual variation in the timing of the dry season and the rains as well as other factors affecting garden yields, such as infestations of locusts and other pests. The magnitude of the average seasonal weight loss among Lese women varies from as much as 5 kilograms to as little as 0.5 kilogram, with the range of individual variations of course being much greater still. A particularly severe hunger season occurred in 1983, for example, following early rains and a very short dry season in the previous year. Lese women lost an average of 8.1 percent of their body weight between December 1982 and June 1983. They successfully recouped their weight loss in the following six months after a good harvest in the summer of 1983. Because of that same good harvest of 1983, the hunger season in the first half of 1984 was relatively mild, with an average weight loss for Lese women of only 2.0 percent. But a poor harvest in June of that year, caused in part by a locust infestation in the spring, led to a continuing pattern of weight loss into the last half of 1984. Another severe hunger season followed in 1985, with an average weight loss of 5.3 percent among the Lese women between December 1984 and June 1985.

As described above, female ovarian function among the Lese declines in parallel with female weight during the hunger season. Estrogen and progesterone levels decline, ovulation is less frequent, cycles are longer, and the period of menstrual flow is shorter. Weights usually begin to increase slowly after the June peanut harvest, and indices of ovarian function begin to improve soon after but may take a month or more to recover significantly. If this seasonal variation in ovarian function reflects seasonal variation in fecundity, we would predict a dearth of conceptions during the hunger season and a peak in conceptions after the harvests, at the height of food availability.

The pattern of births recorded among the Lese from 1980 through 1987 accords perfectly with this prediction. Back-dating births by nine months reveals a statistically significant deficit in conceptions between May and July and a peak in conceptions from September to November. But as pointed out above, the simple fact of a correlation between two seasonal variables does not necessarily imply a causal connection. Two additional observations strengthen this inference in the case of the Lese. Because the Lese do not experience severe hunger seasons every year, we can compare the pattern of conceptions in the years with statistically significant weight loss with the pattern in years when weight changes are not statistically significant. Only in the years with statistically significant weight loss is the seasonal pattern of conceptions also significant. This implies that other seasonally varying factors that do not share the same annual variation as energy balance are unlikely to be causes of the seasonal pattern of Lese births.

A second observation that strengthens the hypothesis that variation in energy balance is responsible for seasonal variation in Lese births concerns the Efe pygmies who live in the same area of the Ituri Forest. The Efe are nomadic hunters and gatherers who trade forest products and labor to the Lese in exchange for garden produce. For much of the year they camp in the vicinity of Lese villages and interact with the Lese on a daily basis. Efe men help Lese men cut and burn garden plots; Efe women help Lese women cultivate and harvest the crops. The Efe participate in virtually all aspects of Lese ritual life, sharing the same puberty and funeral ceremonies, and are even incorporated into the patrilineal descent system that organizes Lese social life. Efe camps are never permanent, however, and the Efe retain the ability to subsist for long periods away from Lese villages, relying primarily on wild game and foraged plant resources. Largely because of this mobility and broader resource base, the Efe are not subject to the same regular pattern of weight loss as the Lese, even though they are exposed to all other seasonal aspects of climate and environment. Nor do they demonstrate a seasonal birth pattern. If anything, there are slightly more Efe children conceived in June and July and slightly fewer in October through December than at other times of the year, exactly the opposite of the Lese pattern, though the differences for the Efe are not statistically significant.

Thus, for the Lese, not only have the individual links in the causal chain connecting seasonality of the environment to seasonality of human births been documented—seasonal rains lead to seasonal harvests, seasonal harvests lead to seasonal food availability, seasonal food availability leads to seasonal energy balance, seasonal energy balance leads to seasonal ovarian function, seasonal ovarian function leads to seasonal conceptions—but the central position of seasonality of energy balance in that causal chain has been confirmed. Births are seasonal only to the extent that energy balance is seasonal, both between and within populations sharing the same environment and the same cycles of culture and ritual. Both among Boston women trying to become pregnant and among Lese women trying to sustain themselves and their families in the tropical rain forest, energy balance appears to influence the probability of conception.

THE FAT BONE CONNECTS TO THE EGG BONE

Full understanding of the relationship between energetic conditions and ovarian function depends on an understanding of the physiological pathways by which the one can affect the other. A number of different pathways have been proposed, and indeed there is no reason why multiple pathways should not exist, either reinforcing each other or mediating slightly different effects. After all, we have seen that energy status, energy balance, and energy flux may have independent relationships to ovarian function. Perhaps information regarding these different aspects of female energetics is carried in different physiological channels.

The most obvious place to look for mechanisms regulating ovarian function is the hypothalamic-pituitary axis. The central nervous system receives a host of inputs from the peripheral regions of the body, and the hypothalamus serves as the way-station connecting the central nervous system to the endocrine system by way of the pituitary. Ovarian function depends on regular, pulsatile release of GnRH and appropriate stimulation by pituitary gonadotropins. There is, in fact, convincing evidence that severe levels of energetic stress are often associated with disruptions of hypothalamic-pituitary activity. This effect is usually manifested as a slowing or cessation of the pulsatile pattern of LH release. Amenorrheic anorexia nervosa patients, for example, have LH profiles that are reminiscent of those of preadolescent girls, without the pronounced hourly pulses that characterize mature gonadotropin production. As menstruation resumes with refeeding and weight gain, LH pulsatility is reestablished. Dramatic reduction of caloric intake, both short-term and long-term, has also been shown to result in interruption or suppression of LH pulsatility. Similarly, female long-distance runners have been observed to have lower LH pulse frequencies and amplitudes than sedentary controls. The functional significance of these subtler changes in LH pulsatility is uncertain, however, since they are not necessarily associated with deficits in ovarian steroid production, rate of follicular growth, or menstrual cycle characteristics.

Several studies, however, demonstrate that suppression of ovarian steroid production can occur in athletes in the absence of alterations in gonadotropin profiles. As we will see below, it also appears that the ovarian suppression associated with hunger seasons among the Lese is not accompanied by altered LH patterns. It is likely, therefore, that ovarian function is sensitive to other physiological signals associated with energetic condition besides variation in hypothalamic-pituitary activity. Some have suggested that differences in energetic condition might be associated with differences in the way ovarian steroids are processed by other body tissues, and that this might indirectly affect fecundity. There is evidence, for example, that adipose tissue contains the enzymes that can convert adrenal androgens into estrogens. Fatter women therefore might produce more of this "extragonadal estrogen" than leaner women do. There is also evidence that body fat may influence the way in which estrogen molecules are metabolized before being cleared in the urine, resulting in metabolites that retain more ability to interact with estrogen receptors.

Changes in peripheral metabolism of these kinds, however, are unlikely to have the expected effects on ovarian function and fecundity. The more extragonadal estrogen produced, for example, the greater the negative feedback on gonadotropin production, leading to a lower level of ovarian stimulation, not a higher one. To fall back on the thermostat and furnace analogy used earlier, increasing the production of extragonadal estrogen is like building a fire in the fireplace of a centrally heated house. To the extent that the fire warms up the house, the thermostat would down-regulate the furnace. Overall circulating levels of heat would remain the same. Circulating levels of estrogens show no relationship to fatness in the normal range of body composition, either. If adipose tissue is making more estrogens in fatter women, the ovary must be making less. Only in obese women does excessive extragonadal estrogen production lead to elevated circulating estrogen levels, and then it is associated with ovarian suppression, even to the point of amenorrhea, rather than superfecundity. If the fire in the fireplace gets too hot, the furnace stops working altogether.

There are, however, even more direct signals of energetic condition that can directly influence ovarian function. Among the most likely candidates are the major metabolic regulators that help control energy availability and tissue proliferation, such as cortisol, insulin, and growth hormone. Cortisol we have already encountered as an important regulator of blood glucose levels. The stimulation of adrenal cortisol production is often accompanied by the suppression of ovarian steroid production. Conditions that lead to cortisol elevation, such as high energy expenditure or low energy intake, could therefore result in some degree of ovarian suppression. Ordinarily, cortisol elevations are transitory events that occur in response to acute circumstances. However, chronically low energy intake can be associated with chronic cortisol elevation. High levels of psychological stress are also often associated with elevated cortisol, as the body activates the mechanisms for rapid energy mobilization typical of emergency responses. If severe enough or sustained enough, such a state of "false" energy demand may also be associated with disruption of reproductive function.

Insulin is another hormone that helps regulate blood glucose, but generally in the direction opposite to that of cortisol. Insulin is elevated under conditions of positive energy balance and stimulates the incorporation of glucose either into fat tissue as stored energy or into additional muscle tissue as increased lean body mass. Elevations of insulin occur as transitory events after a meal to stimulate tissue uptake of glucose. Individuals that consume more calories than they expend on a regular basis, however, tend to have higher baseline insulin levels as well, while athletes and others who tend to have high energy expenditure have lower baseline insulin levels. Insulin thus seems to be sensitive to both energy balance and energy flux.

In addition to its effects on fat and muscle and other peripheral tissues, insulin stimulates ovarian cell proliferation and promotes steroid production. Physiological levels of insulin have been found to stimulate the production of both estradiol and progesterone by human granulosa cells in vitro. It has also been shown that this effect is mediated by insulin receptors on those cells, and not by inappropriate stimulation of receptors for the separate but structurally similar insulin-like growth factor I (IGF-I). Insulin acts synergistically with FSH and LH, so that the amount of steroid produced for a given level of gonadotropin stimulation is increased.

Growth hormone is a pituitary protein hormone that, as its name suggests, plays a central role in regulating somatic growth during childhood and adolescence, particularly in stimulating epiphyseal growth of the long bones and the development of lean body mass. Many of growth hormone's effects are mediated or augmented by the actions of IGF-I, a hormone that is produced locally in target tissues. Recent work has demonstrated that growth hormone directly stimulates ovarian production of both estradiol and progesterone. In vitro experiments have demonstrated that physiological levels of growth hormone produce dose-dependent variation in estradiol production and follicular maturation, even in the absence of gonadotropins. A direct ovarian effect of growth hormone independent of gonadotropin mediation is also supported by demonstration of growth hormone receptors in granulosa cells and in the corpus luteum. IGF-I has also been shown to stimulate ovarian production of both estradiol and progesterone and granulosa cell proliferation. IGF-I receptors have also been identified in human ovarian tissue, particularly in the corpus luteum. Comparison of IGF-I concentrations in the ovarian vein emanating from the ovary containing the corpus luteum with concentrations in the ovarian vein from the other ovary indicate that IGF-I is produced within the ovary itself, although the contribution to overall circulating levels is slight. As in other tissues, it appears that IGF-I may be produced locally in ovarian tissue in response to growth hormone, regulating the proliferation of the highly mitotic granulosa cells of both the developing follicle and the corpus luteum and thereby affecting the steroid production of those same tissues.

The growth hormone/IGF-I system broadly regulates many anabolic and tissue proliferative processes in the body, and not surprisingly its activity is sensitive to the constraints of energy availability. Short-term fasts and exercise increase growth hormone levels but depress IGF-I levels. Chronic undernutrition, by contrast, is associated with down-regulation of the whole system and lower levels of both growth hormone and IGF-I. The granulosa cells of the developing follicle and developing corpus luteum are among the most mitotically active of normal human tissues, and the steroid production of the follicle and corpus luteum is generally proportional to the degree of granulosa cell proliferation. The effect of growth hormone and IGF-I on ovarian steroid production may thus be mediated in two ways: by stimulating increased granulosa cell proliferation and by stimulating increased steroid production.

Leptin is one of the newest players on the endocrine stage that may also adjust ovarian function in relation to energetic state. Leptin is a peptide (a short chain of amino acids) produced by fat tissue that appears to help control feeding behavior. Mice, or people, deficient in leptin because of genetic defects will eat to the point of morbid obesity. Administering leptin effectively curtails this overeating and leads to weight loss. Unfortunately, the visions of a wonder drug for painless weight loss that danced in the heads of pharmaceutical company executives when leptin was discovered have not been realized. In normal humans, who do produce leptin naturally, extra leptin administered through a syringe doesn't appear to have much effect on eating or weight gain.

Leptin does appear to reflect all three dimensions of energetic condition that we have identified, however. In individuals at neutral energy balance it varies with fat mass, hence with energy status. In individuals who are gaining or losing weight, the rate of leptin production per gram of fat rises or falls respectively. Finally, when subjects' weights are experimentally raised or lowered and then maintained at the new level—at a new level of energy flux but neutral energy balance—the new rate of leptin production per fat mass is maintained as well. Insulin acts as a direct stimulant of leptin production by fat cells, so it may be that variation in leptin production is jointly reflecting fat mass and insulin dynamics.

Most tantalizing are suggestions that leptin may influence ovarian function. In mice, manipulating leptin levels in immature individuals can accelerate or retard reproductive maturation. In girls, leptin levels normally rise during adolescence. The ovary appears to possess leptin receptors. Leptin levels appear to change during the menstrual cycle in association with estrogen levels. However, in most of the associations documented for humans it is difficult to determine the direction of causation. Increasing leptin during adolescence may be a result of increasing fatness, hence of increasing ovarian function. Similarly, variation in leptin levels across the menstrual cycle can reflect estradiol's enhancement of insulin's effect on fat cell function. Both leptin levels and leptin effects on the ovary appear to be mediated by insulin. Thus rather than being an independent factor linking energetic variables to ovarian function, leptin may be part of a single system linking metabolic state and reproductive function. If leptin does turn out to have an effect on ovarian function, it will largely parallel the effect of insulin and hence be a redundant and reinforcing signal of energetic condition.

Two basic pathways thus exist for the regulation of ovarian function: central regulation via the hypothalamus and pituitary, and peripheral regulation via the major regulators of energy metabolism. Determining which of these pathways is responsible for ovarian suppression in any given circumstance is difficult, however. Obtaining evidence of altered LH pulsatility usually requires repeated blood sampling several times an hour for extended periods. A reflection of this pulsatile LH activity can be observed in progesterone patterns. As noted earlier, progesterone is produced by both luteinized theca cells and luteinized granulosa cells, but only the luteinized theca cells respond to LH pulses with pulsatile release of progesterone. Progesterone pulses can be observed and measured (if samples are collected frequently enough) in both plasma and saliva with comparable results in the two media. The early luteal phase is characterized by pulses that are low in frequency and low in amplitude. The frequency of progesterone pulses increases in the middle of the luteal phase to between 0.5 and 1.0 pulses per hour and then declines in the late luteal phase.

My colleagues and I have collected saliva samples at fifteen-minute intervals for periods up to eight hours from recreational runners and controls in Boston and from Lese women in the Ituri Forest. In the Boston sample the runners displayed progesterone pulses in the same frequency and amplitude as the controls, even though their baseline levels of progesterone were lower. Similarly, among Lese women weight loss during the hunger season was associated with a decrease in baseline progesterone levels, but there was no change in the frequency or amplitude of progesterone pulses. These results indicate that in both cases the suppression of ovarian function observed was not mediated by changes in LH pulsatility. Pulsatile production of progesterone by the luteinized theca cells appeared to be unaffected. Rather it appeared to be the tonic production of progesterone by the luteinized granulosa cells that was suppressed. This result would be consistent with regulation via the peripheral pathway resulting either in reduced proliferation of the granulosa cell mass in the developing follicle, or in lower stimulation of progesterone production by the luteinized granulosa cells, or both.

Coupling ovarian steroid production to energy availability is thus not a difficult physiological challenge for natural selection to meet. All that is required is the expression of genes promoting the receptors for hormonal signals already in place in a manner comparable to that of other proliferating tissues. The result is a system of quantitative, rheostat-like regulation of ovarian function distinct from the system of qualitative, on-off regulation of ovarian function localized in the hypothalamic-pituitary axis. This two-stage regulatory system is capable of modulating ovarian function along its continuum in response to variable energetic stress in exactly the ways we have seen demonstrated. The peripheral pathway of ovarian regulation in response to energetic stress also seems more efficient than the central pathway. Rather than requiring a message regarding energetic condition to be sent from the body to the brain, triggering another message to be sent from the brain to the ovary, the ovary can receive the message of energetic condition directly. During periods of extreme stress, and at extremely low levels of energy status, the hypothalamic-pituitary-ovarian axis seems to be turned off at the top. But there is also abundant evidence that ovarian function adjusts in a more quantitative, rheostat-like way to energy balance and energy flux across a range of conditions normally encountered by human beings. Finally, evidence indicates that variation along the continuum of ovarian function translates into variation in female fecundity.

THE WISDOM OF THE BODY

The question remaining is "why?" Why should natural selection have designed ovarian function to respond in this way? The evidence indicates two important differences from the relationship between fatness and fecundity originally envisioned by Frisch. First, fatness per se doesn't seem to be the only, or even the key, aspect of a woman's energetic condition to which her reproductive physiology responds. As long as a woman isn't emaciated or obese, ovarian function seems to be more sensitive to normal variation in energy balance and energy flux than normal variation in stored fat. Second, the female reproductive system doesn't respond to variation in energetic condition simply by turning on or off. Rather than a threshold, there is a quantitative continuum of response. Greater energetic stress elicits greater ovarian suppression, and lesser stress lesser suppression. We can direct our question "why" at both of these differences.

Why should ovarian function be sensitive to energy balance and energy flux independent of energy status? This would be logical if energy balance and energy flux provide independent information regarding the chances of reproductive success that simple fatness does not convey. Successful reproduction, as we have observed, depends on a woman's ability to allocate energy to that end over and above her own maintenance needs. She must start to allocate such surplus energy immediately upon conception, even though the energetic demands of the pregnancy are negligible at that point, in order to meet the very high demands of late gestation and unsupplemented lactation. Energy stored as fat before conception could theoretically be used for gestation, but empirically it is not. Postpartum weights are higher than preconception weights even in places like the Gambia and Bangladesh. Even if stored fat was directly converted into baby, it is very unlikely that it could, by itself, meet the demands of the pregnancy without endangering the life of the mother. Human reproduction does not involve women accumulating fat to some critical level, converting it all into a baby, and then starting the cycle again. The fat that women carry on their bodies in a nonpregnant, nonlactating state serves principally to buffer their own survival against periods of negative energy balance, the same function that it serves in men. If the buffer gets so low that survival is threatened, nonessential energy allocations are suspended, including reproduction. Short of such emergency situations, some of the buffer might be redirected toward reproduction in times of need, but the real key to successfully meeting the energetic demands of pregnancy and lactation is the ability to generate surplus energy over maintenance requirements on an extended basis.

If surplus energy is the key, then positive energy balance is clearly an important predictor of success. Positive energy balance results from an energy intake that exceeds expenditure. Just as clearly, negative energy balance lowers the probability of success, since it indicates that current intake is insufficient to meet current expenditure, much less the added cost of pregnancy. As we noted above, energy balance is independent of energy status. Two women can have the same level of fatness and yet differ in their energy balance. If ovarian function were designed only to be sensitive to energy status, it would not be able to distinguish between these conditions. Being sensitive to energy balance in addition to energy status allows for an increased probability of conception when the likelihood of surplus energy is greatest.

The reasons for ovarian sensitivity to energy flux are less obvious. Grazyna Jasienska has suggested that high energy flux may ordinarily have been associated with negative energy balance in the formative human past, since high energy expenditure would have been more likely to occur in pursuit of scarce calories than as a reaction to caloric abundance. But that does not explain the function of a separate sensitivity to energy flux. Sensitivity to energy balance alone would suffice. It is likely, however, that variation in energy flux independent of energy balance may constrain a woman's ability to generate the surplus energy needed for reproduction. A woman in neutral energy balance but high energy flux is already allocating more than the ordinary number of calories to her own maintenance requirements. Her ability to generate additional surplus energy above these high maintenance requirements may be less than the ability of a woman in more moderate energy flux. (This assumes that the high energy expenditure is ecologically necessary, and not an idiosyncratic and voluntary expenditure. Presumably our Pleistocene ancestors did not go jogging for health.) Similarly, neutral energy balance together with low energy flux would indicate a condition in which expenditure on maintenance must already be lower than normal to avoid negative energy balance. Again, the likelihood that surplus energy can be generated under these conditions is reduced. (This assumes that the low energy intake is ecologically dictated and not idiosyncratic and voluntary. Presumably our Pleistocene ancestors didn't diet for looks either.)

By this logic, energy status alone is a poor predictor of the ability to generate surplus energy in the future since it doesn't even indicate that ability in the present but only points to the ability some time in the past. Energy balance and energy flux are better indicators of present ability, the ability that will be called upon even in the first weeks and months of pregnancy. Even these variables do not predict ability to generate surplus energy in the future, however. If humans evolved in very predictable environments where surplus energy availability could be anticipated confidently from environmental signals—day length, temperature, rainfall—then we might expect ovarian function to key on these variables rather than on indicators of current energetic conditions. Many high-latitude species of land vertebrates use seasonal cues to coordinate their reproductive systems. There is less reason to think that humans would have evolved this ability, however, given the evidence of our recent African origins. Nor does there seem to be much evidence of direct climatological sensitivity in humans. Efforts to find evidence for an influence of photoperiod or temperature on human fecundity have largely been unsuccessful and have never claimed a strong influence in any event. If there are no good cues of energetic conditions to come, current energetic conditions become the best variables to use to optimize the chances of meeting the energetic demands of reproduction.

Given that the best indicators of the ability to generate surplus energy for reproduction are energy balance and energy flux, why does the female reproductive system respond to them in a continuous fashion rather than simply turning on or off? "On or off" regulation might make sense if any given moment could be categorized as either a "right" or a "wrong" time for conception. If this is a "wrong" time to conceive (meaning the probability of success is low), a woman would do better to wait for a "right" time (probability of success high). Conceiving at the "wrong" time results only in a waste of time and energy, possibly to the detriment of the woman's own chance of survival.

But classifying all opportunities for reproduction as "right" or "wrong" times in this way does not seem like a good algorithm for optimizing reproductive success. It doesn't take into account the magnitude of difference between "right" and "wrong" times in the chances of success or the length of the wait necessary for a "wrong" time to be succeeded by a "right" time. If a relatively short wait would result in a large improvement in the chances of success, then the wait might be worth it. If, on the other hand, a very long wait was necessary to achieve only a marginal improvement in the probability of success, it might not be justified. Nor does the "right time / wrong time" approach take into account the confidence of the categorization. This seems particularly important given our appreciation of the fact that future energetic conditions may not be predictable. Suppose that instead of a dichotomous categorization system of "right" times and "wrong" times we can only place opportunities to reproduce along a continuous scale, from very low probability of success to very high probability of success. Then the optimal allocation of limiting resources (time and energy) when the future is unpredictable would be based on scaling the investment to the probability of success, not on responding to variation in conditions along this continuum in an "all-or-nothing" manner. In addition, allowing ovarian function to vary on a quantitative scale keeps the door of opportunity open for superior embryos whose appearance is also unpredictable.

Knowledge of the future, even partial knowledge, would change the optimal strategy. To the extent that it is "known" that future conditions will not improve, the value of waiting over trying to reproduce now, even if conditions are poor, is reduced. But that does not necessarily mean that the optimal allocation of limited resources under chronically poor conditions is the same as it would be under chronically good conditions. Waiting longer to conceive may have intrinsic value to a woman living under poor energetic conditions over and above the possibility of conditions improving. Remember that given the high energy demands of pregnancy and lactation (metabolizing for two), the waiting time to conception can represent a critical period of energetic recovery (metabolizing for one). Especially under chronically poor energetic conditions, a woman's lifetime reproductive success may depend on her ability to maintain long-term energy balance and avoid the downward spiral of maternal depletion. The waiting time to conception may be her only opportunity to recover from the energetic deficits incurred during pregnancy and lactation. Lowering her probability of conception is one way to achieve this extension of the respite of metabolizing for one. Even a modest increase in waiting time of a few months can represent a large proportional increase in the time spent in this state.

Hence we might expect that women living under conditions of chronic energy shortage would have lower average levels of ovarian function than would women living under chronic energy abundance. One sign of chronic energy shortage, as we saw in the last chapter, is slow growth and late maturation. There is, in fact, some evidence that late maturation is associated with low ovarian function throughout adulthood. Longitudinal studies in Finland have indicated that late maturing girls have lower steroid levels and lower frequencies of ovulation than early maturing peers well into their twenties. Cross-sectional studies in Italy indicate that later maturing women have higher frequencies of ovarian dysfunction as adults than earlier maturing women do. Studies of ovarian function among various populations around the world have demonstrated a significant negative correlation between average menarcheal age and average luteal progesterone level among women twenty-five to thirty-five years old. This pattern fits Eric Charnov's model of growth and reproduction, discussed in the previous chapter. According to that model, both growth and reproduction represent the investment of residual metabolic energy over and above the costs of maintenance. In environments where such residual energy can be generated only at a slow rate, both growth rate and reproductive rate are expected to be slow as well. Lowering fecundity and extending the waiting time to the next conception provides a mechanism for females to achieve a lower reproductive rate and maintain longterm energy balance.

It seems that ovarian function in women has been artfully designed by natural selection to make the best use of the two primary limiting resources for female reproductive success: time and energy. But such modulation is not only a matter of adjusting a lifetime baseline for ovarian function and responding to variation in energetic conditions with short-term adjustments relative to that baseline. The baseline itself changes throughout life in a characteristic fashion. This arc of reproductive life and the forces that have shaped it are the subject we turn to next.

THE ARC OF LIFE

IN THE FEBRUARY 18, 1982, issue of *The New England Journal of Medicine,* French researchers presented data for over 2,000 French women whose husbands had been diagnosed as azoospermatic, incapable of producing viable sperm, and their efforts to conceive by artificial insemination with donor semen. They reported that the success rate for this procedure dropped with the age of the woman, beginning as early as age thirty and accelerating after age thirty-five. Under the controlled circumstances of these treatments, neither the age of the husband nor changes in the frequency of intercourse could be invoked as possible explanations of the result. An accompanying editorial noted: "Individual and societal goals may . . . have to be reevaluated. Perhaps the third decade [of a woman's life] should be devoted to childbearing and the fourth to career development, rather than the converse, which is true for many women today."

Dozens of articles rapidly appeared in the popular press in the wake of the French report repeating and magnifying the alarming message that women who choose to delay childbearing might be at risk of forgoing it entirely. The "biological clock" had begun to sound in the ears of the public, and to many its ticking seemed ominously loud.

In the October 25, 1990, issue of *The New England Journal of Medicine,* Mark Sauer, Richard Paulson, and Rogerio Lobo of the University of Southern California seemed to turn off the alarm on the biological clock. They reported achieving successful pregnancies in five of seven women who had already been through menopause, using eggs from younger donors, fertilized in vitro, and introduced into the mother's uterus. Endometrial development in the mother had been artificially stimulated and synchronized with follicular development in the donor by the administration of exogenous steroids mimicking a natural conception cycle. An accompanying editorial noted, "We now have extraordinary control over a process that was once inexorable in its demands and limitations, particularly for women." H. Thomas Murray of Case Western Reserve Medical School was quoted in the *Boston Globe* a few days later, commenting on the reaction among both laypeople and clinicians to the report by Sauer and his colleagues: "This is in some ways more startling and threatening than almost all the other reproductive technologies except for surrogacy . . . I think it's because it violates our understanding of the natural stages of life. It threatens what we have come to rely upon as a predictable transition in life for women . . . an age that has always been a reliable milepost in the march through life. And now the milepost has been ripped out and carried an indeterminable distance, and nobody knows where it's going to be put down again."

Progress in understanding and treating infertility seems to cut two ways. On the one hand it reveals the inexorable nature of the physiological limitations on female fecundity, and on the other hand it offers solutions that seem to free women from the tyranny of "natural stages of life." The two cuts of this sword are, of course, related. The demand for infertility treatments of an increasingly complex and technological sort comes in large part from older couples who have, for one reason or other, deferred starting a family until late in the woman's reproductive life. Understanding the natural arc of female fecundity has come slowly and is as yet incomplete. But what we do know suggests that the relationship of age and female fecundity is a basic part of human biology, shaped by forces of natural selection that are both very old and possibly, on an evolutionary time scale, quite new.

WHAT GOES UP MUST COME DOWN

In Louis Henry's original analysis of natural fertility there are two remarkable results. Earlier we stressed the sizable variance in *levels* of natural fertility exhibited by different populations and the role of lactation in explaining those differences. Equally notable, however, is the similarity in the *shape* of natural fertility by age across populations. In fact, Henry found that the curves of age-specific, marital female fertility for different populations were extremely close: virtually superimposable if they were all expressed as percentages of the fertility of women twenty to twenty-four years old. Henry probed this observation further to inquire whether some common risk of sterility could account for the similarity in pattern across such a wide range of levels of fertility. He did this by calculating the proportion of couples at each age that clearly remained fecund, as evidenced by the subsequent birth of a child. This proportion did indeed decline along a rather uniform trajectory, though not without some significant differences between populations. The decline in the percentage of fecund couples, however, was not of sufficient magnitude to account for the entire decline in fertility that accompanied increasing female age. When the age-specific fertility rate of those women who did go on to the next level of parity was examined, Henry found that it, too, declined along a uniform trajectory that was very close to that of the proportion of fecund couples. He concluded that "the decrease in the fertility of all couples could be attributed to two factors of more or less equal importance: the reduction in the percentage of fecund couples and the reduction in the fertility of these couples." The close correlation between the two curves also suggests that the two processes are not unrelated. Nor does controlling for the husband's age eliminate the trend. Although it may be possible to account for this fact through other mechanisms, a steady decline in the physiological basis of female fecundity with increasing age is an almost unavoidable hypothesis.

The most important competing hypothesis for the decline in age-specific fertility across populations is that it is a consequence of declining frequency of intercourse with age. Several empirical studies have documented that the frequency of intercourse within marriage declines with the age of both husband and wife and with the duration of the marriage. This pattern of decline is quite similar in different populations, despite variation in cultural norms. Could waning of sexual activity by itself be responsible for the regular decline in natural fertility with age?

Maxine Weinstein and her colleagues compared the effect of age-related changes in the frequency of intercourse and the effect of age-related changes in female fecundity as competing explanations for age-related changes in natural female fertility. The data on coital frequency suggest a pattern of steady decline with male age, female age, and marital duration. Such a pattern doesn't help at all to explain the increase in marital fertility with age before age twenty to twenty-four. Nor does the linear decline in coital frequency seem to account for the accelerating decline in natural fertility with advancing age. The data on both early embryonic loss and ovulatory frequency, on the other hand, present U-shaped or J-shaped (inverse for ovulatory frequency) relationships with female age. These patterns of age-related change in female fecundity are compatible with fertility rates that are low immediately after menarche, rise during adolescence to a peak in the twenties, and decline at ever increasing rates after age thirty. Hence the observed pattern of age-specific natural fertility is unlikely to result from changes in coital frequency and much more likely to result from changes in female reproductive physiology.

The most convincing evidence of age-specific changes in female fecundity comes from the realm of "assisted reproductive technologies," or ART. As noted above, variation in coital frequency cannot explain away the results of the French artificial insemination study because normal intercourse was not involved and all inseminations were timed to the woman's ovarian cycle. Similar levels of control are characteristic of most ART. If fertilization is not completely in the control of the physicians, the timing of intercourse is prescribed. Age-related declines in pregnancy rates have now been documented in almost every ART procedure, including ovulation induction and in vitro fertilization. Indeed, so common and well documented is the effect of female age that its absence in any substantial database would be a significant, publishable finding. Further, there is good reason to suspect that a large portion of the age-specific change in female fecundity involves changes in ovarian function. The successful pregnancies that Sauer and others have achieved in older women using donor eggs and exogenous stimulation of the uterus suggest that the uterus retains its functional capacity at an advanced age *if* appropriate steroid support is supplied. Instead of a nonfunctioning uterus, it seems to be a deficiency in one or both aspects of ovarian function—gamete maturation and steroid production—that is limiting on female fertility at older ages. The rise in age-specific female fertility before the mid-twenties in natural fertility populations is likely also to be due to positive changes in one or both of these aspects of ovarian activity. Thus by tracing the rise and fall of ovarian function across the female reproductive life-span we may be able to uncover the biological foundation for the characteristic age-specific pattern of natural human fertility.

THE UPWARD PATH

There are two prevalent misconceptions, one popular, the other quasi-scientific, regarding female fertility in the years following menarche. On the one hand are cultural messages celebrating youthful vitality laden with images conveying the sexual attractiveness of women in their late teens. These subliminal messages often lead us to assume that a woman is at her peak of fecundity as she enters adulthood. On the other hand is the notion of "adolescent sterility," a post-menarcheal refractory period during which a woman is supposed not to be capable of pregnancy. This idea, promoted in an early monograph by Ashley Montagu, gained currency for a while in certain academic fields, including demography. As it turns out, neither of these representations of adolescent fecundity is correct. Instead, ovarian function traces a steady and sustained rise in the decade after menarche even as it traces a steady decline in the decade preceding menopause. At the extremes of reproductive age, changes in menstrual patterns make the existence of underlying physiological changes more apparent. Menstrual cycles are often irregular in the years immediately after menarche or immediately before menopause. But an appreciation of the more sustained dynamism of ovarian function across a woman's reproductive life depends on a closer observation of hormonal patterns.

One of the earliest studies of variation in human ovarian function with age was carried out by Gerhard Döring and published in 1969. Döring collected records of basal body temperature from 481 German women over a total of 3,264 menstrual cycles. Because progesterone causes an elevation of basal metabolism, changes in basal body temperature recorded daily and charted over the course of a month can be used to infer ovulation and luteal activity. This method is fraught with difficulties and potential sources of error, but in experienced hands can provide a good deal of information. The relationship between basal temperature elevation and circulating progesterone is qualitative, not quantitative. One can infer the presence of an active corpus luteum and some measure of the length of time that luteal activity persists, but not the level of activity attained. Döring was able to use his data to characterize cycles as ovulatory or anovulatory on the basis of the presence of a basal temperature rise, and was further able to categorize ovulatory cycles as "luteally sufficient" or not on the basis of the length of time that basal temperature remained elevated. When the frequencies of these categories of ovarian function were plotted against age, the first full picture of the rise and fall of female ovarian function was produced.

As limited in individual quality as these data are, in aggregate they represent a data set that is remarkably rich and robust. In the histogram Döring constructed, one can observe many of the important features of age-related changes in ovarian function that have subsequently been confirmed by more recent studies. Particularly striking are the long sustained rise in ovarian function in the first years after menarche, a rise that does not culminate until the mid-twenties, and the decline in ovarian function that is under way by age forty. The decline in ovarian function at older ages is less steep than the increase in ovarian function at young ages. It should be noted, however, that the data are representative only of menstruating women. The decline in ovarian function with advancing age beyond forty might appear more striking if women experiencing irregular cycles were included in the sample. Furthermore, one must remember that the data involved are cross-sectional, not longitudinal. Individual trajectories may vary quite widely.

The "arc" of ovarian function that Döring described has been substantially supported in more recent and more probing studies. For example, Susan Lipson and I measured salivary progesterone in daily samples collected by over 150 regularly menstruating women between the ages of eighteen and forty-five to provide a more direct and more quantitative description of age-related changes in ovarian function than Döring's. The composite profiles generated from these data demonstrate in graphic form the rise and fall of ovarian function with age. Similarly, different quantitative indices of ovarian function, such as ovulatory frequency, or average progesterone production, trace comparable parabolic trajectories against age. In nearly every case, however, the rise at young ages is steeper than the decline at older ages.

Dan Apter, Reijo Vihko, and their colleagues have documented several important characteristics of adolescent ovarian cycles that distinguish them from the cycles of older women. Cycles in the teenage years tend to have high profiles of testosterone and other androgens, especially in proportion to the levels of estradiol and other estrogens. Follicular maturation, as reflected in the growth of individual follicles measured ultrasonically over the first half of the cycle, is slower than in older women, and the ultimate size of the ovulatory follicle, when there is one, is smaller. The amount of progesterone secreted by the corpus luteum, as we have already seen, is smaller than in older women and is correlated with the size of the immediately preovulatory follicle. All of these characteristics suggest that it may be the developmental capacity of the follicle that continues to mature after menarche. The high androgen to estrogen ratio of adolescent cycles suggests that the granulosa cells in young women are not as efficient at aromatizing testosterone as the same cells in older women. This may well be a consequence of the smaller size of the adolescent follicle. Certainly, small follicular size seems related to a diminished capacity to produce progesterone after ovulation. The smaller size of the preovulatory follicle, if it results in lower rates of estradiol production, may also result in an ovum of diminished fertilizability, since the ability of the ovum to play its part in the chromosomal dance of zygote formation depends on exposure to a sufficient estradiol level during its final development.

It's possible that all these differences could be the result of lower levels of pituitary stimulation in younger women: It may not be ovarian function per se that increases during adolescence but rather pituitary function. Some evidence of this can be derived from studies of ultradian patterns of progesterone production during the luteal phase of the cycle. In my laboratory my colleagues and I have compared the pattern of progesterone pulses in samples of saliva collected every fifteen minutes for eight waking hours from Boston women eighteen to twenty-two years of age and women twenty-three to thirty-five years of age. The comparison was made both in the middle of the luteal phase, when average progesterone levels are at their peak, and in the last week of the luteal phase, when the ability to respond to hCG in the case of fertilization may be crucial to the success of a pregnancy. At both points in the cycle the younger women have both fewer and smaller progesterone pulses than the older women. Baseline progesterone levels are also lower, suggesting that progesterone production by both luteal cell populations may be reduced in the women eighteen to twenty-two years old relative to the women twenty-three to thirty-five years old. As we will see, this contrasts with the situation observed in women over thirty-five. Differences in baseline production and in pulse amplitude could well reflect differences in the capacity of the corpus luteum to produce progesterone. But differences in pulse frequency are most often assumed to reflect differences in LH pulse frequency, and hence differences in pituitary activity.

In every measurable aspect of ovarian function, then, women in their late teens and early twenties seem to come out lower than women five to ten years older and to represent not a peak of reproductive capacity but an intermediate point on a steadily rising trajectory. Nor is this pattern limited to one particular population of women. Similar age patterns can be observed among the Lese of the Congo's Ituri Forest and the Tamang of Nepal. Genetically, geographically, culturally, and ecologically these two populations are extremely different, both from each other and from the Boston women discussed above. The average levels of salivary progesterone that we observe in both the Lese and the Tamang are also much lower than those observed in Boston. Yet in both populations women in their late teens to early twenties show lower indices of ovarian function than do women in their late twenties and early thirties. Indeed, the ratio of progesterone indices between these two age groups is quite similar across all three populations.

Both the data on natural fertility and the data on ovarian function seem to be consistent with the notion of a steady, age-related increase in fecundity over the decade or so after menarche. These observations, however, seem to beg a theoretical question. Can this pattern of maturation be understood as the product of evolution by natural selection? Once the necessary constraints of physical size are overcome, wouldn't it be to a woman's individual selective advantage to rapidly attain peak fecundity and then to sustain it as long as possible? Wouldn't the predicted "optimal" curve of age-specific fecundity be nearly rectangular rather than the observed parabolic trajectory of ovarian function? Here we must remember that the pattern of rising fecundity at young ages may have causes and explanations that are distinct and unrelated to the pattern of declining fecundity at older ages.

Two considerations may help to account for the steady rise of female fecundity at young ages. First is the evidence that young girls themselves do not fully complete their own processes of physical maturation until the late teens or early twenties. Complete fusion of the epiphyses of the long bones is ordinarily not complete until an average age of eighteen years in American and British girls. The same is also true for changes in body composition, breast development, muscle mass, and other aspects of soft tissue development. And pelvic remodeling is characteristically the very last phase of skeletal development in girls, again not completed until the late teens in American girls. There appear to be significant costs associated with reproducing while in an immature state. Rates of infant and maternal mortality are both high for young adolescent mothers and rise steeply with decreasing maternal age. The frequency of low birth weight infants is also high among young mothers, with respect both to absolute age and to years since menarche, or gynecological age. Thus, insofar as a woman's own incomplete development makes reproduction riskier, or places her in more intense metabolic competition with her own fetus, we might expect that natural selection would have acted to modulate relative fecundity. Or one might turn the perspective around and think of the intense selection pressure favoring early reproduction as responsible for a positive level of fecundity in young women who are not yet fully mature.

A second rationale for increasing fecundity at young ages follows from more formal models of life history phenomena. Theoretically, each female faces a trade-off between reproduction and her own survival. The high rates of infant and maternal mortality referred to above translate into a high cost/benefit ratio for reproduction in very young women relative to what they can expect in a few years. However, in theoretical terms risks associated with reproduction are heavily weighted by a woman's expectation of future fertility. In accepting a certain level of risk a young woman places in jeopardy a lifetime of potential childbearing. An older woman jeopardizes only her remaining reproductive career. This measure of expected future offspring, known as reproductive value, declines progressively from the onset of reproductive maturation as a woman's opportunity for further reproduction slips away, leading to the prediction that reproductive effort (in the physiological, not the behavioral sense: the metabolic commitment to reproduction) should increase with age even if fertility and mortality probabilities are constant. According to this prediction, we ought to see female physiology weighted more and more heavily toward reproduction with each passing year after menarche as the fitness costs of increased mortality risk become less and less. From this perspective the mystery is not why female fecundity continues to increase in the twenties, but why it does not continue to increase throughout a woman's life!

THE FULLNESS OF TIME

The decade or so from the mid-twenties to the mid-thirties represents something of a plateau in ovarian function, at least as depicted by cross-sectional data. There is a tendency toward slightly shorter menstrual cycles, but nothing to write home about. The absence of strong age effects should not, however, be misconstrued to mean that that there is little variability in ovarian function during this period. As we have already seen, pregnancy and lactation can cause an individual woman's fecundity to drop repeatedly to zero, while variations in energetic condition can have similar if more moderate effects. Equally impressive, yet masked to a degree by cross-sectional data, is the high degree of variability between women in hormonal profiles and in the frequency and regularity of menstruation and ovulation.

Failure to appreciate normal variability in ovarian function may underlie certain clinical dilemmas, such as the difficulty in establishing clear diagnostic criteria for luteal phase deficiency, or LPD. Marguerite McNeeley and Michael Soules reviewed the difficulty that physicians face in trying to identify LPD, a clinical syndrome of frequent spontaneous abortion, often early in pregnancy, associated with insufficient or out of phase endometrial maturation. Numerous diagnostic approaches have been attempted, based on basal body temperature charts, single serum progesterone determinations, timed and/or multiple serum progesterone determinations, and endometrial biopsies. None of the measures proposed achieves an acceptable discrimination of LPD patients from the unaffected population. To make matters worse, diagnostic classifications don't stick. Women classified in one group this month may fall in another group next month.

In fact, this difficulty may derive from a perspective on normal biological variation that is peculiar to clinical medicine, a dichotomous view of the world that equates "normal" with "healthy" or "free from pathology," a state that is ultimately distinct from the alternative "abnormal," "diseased," or "pathological" state. Clinical decisions are based on discriminating between these two alternative states of the organism, and diagnostic criteria are evaluated in terms of their ability to make such discriminations. LPD, however, may represent a case where this model breaks down entirely. Rather than representing a bimodal distribution of women along some diagnostic measure, the presumed pathology may simply represent a relative position on a single continuum of ovarian function, all of which is "normal" in the sense of the epidemiologist or the human biologist. What is "low" progesterone for one woman may be "high" for another, and an individual woman may move up or down on the continuum of ovarian function either idiosyncratically or in response to any number of behavioral, constitutional, or ecological factors. An appreciation of this degree of normal variation in ovarian function might lead to a different conception of the problem of LPD and its treatment.

Unfortunately, both our appreciation of variability in ovarian function in mid-reproductive life and our understanding of declining ovarian function preceding menopause are severely handicapped by the absence of good longitudinal data, so that our conclusions about changes in ovarian function in this period must be very tentatively drawn. The general picture that emerges from cross-sectional data is one of gradual decline in ovarian function during late reproductive life, detectable as early as the late thirties and accelerating during the forties. This is apparent both in Döring's original data and in the progesterone profiles of Boston women presented above. Salivary estradiol levels decline with advancing age after age thirty in a pattern somewhat different from that shown by progesterone levels. Aligned relative to the midcycle peak of estradiol, both follicular and luteal levels are lower in older women. Midcycle peak levels do not differ with age, as might be expected if the final size of the preovulatory follicle does not change. The length of the follicular phase increases, however, indicating that it takes a longer time for a given dominant follicle to achieve a preovulatory state in older women. The length of the luteal phase, by contrast, is shorter in older women as overall cycle length contracts slightly. If anything, the decline in estradiol levels is significant at an earlier age than the decline in progesterone levels. In this regard it is interesting to remember that estradiol production depends on the capacity of follicular granulosa cells to aromatize androgen precursors, and that granulosa cells are as old as the oocyte they enclose, having remained in "suspended animation" along with the oocyte from the time of follicle formation in the second trimester of fetal life. Theca cells associated with the follicle, by contrast, are newly differentiated from the ovarian matrix. Eggs and granulosa cells may thus both be "old" in comparison with "new" theca cells.

Observations made on patterns of pulsatile progesterone in older women further differentiate the pattern of age-related changes in progesterone from that shown by changes in estradiol. There is an increase, rather than a decrease, in the frequency and amplitude of progesterone pulses observed in saliva samples of women over thirty-five compared to women twenty-five to thirty-four years old. This increased pulsatile production on the part of luteinized theca cells may in part compensate for diminished tonic secretion by luteinized granulosa cells and so act to help sustain the average progesterone levels of older women. Sustaining average levels of luteal progesterone may be important for adequately preparing the endometrium for implantation. Even more crucial may be the ability of the corpus luteum to respond to hCG. In this regard the relative "youth" of the theca cells and their robust pattern of pulsatile progesterone production in the luteal phase may be important in sustaining the fecundity of older women. There may also be a mechanism linking lower estradiol levels in the aging ovary to increased pulsatile progesterone production. Lower luteal estradiol has been shown to be associated with increased release of pituitary LH, presumably because of a lower level of negative feedback restraint, and increased pituitary LH release may lead to increased progesterone production on the part of the responsive luteinized theca cells. Thus one aspect of declining granulosa cell function (lower estradiol levels) may produce a beneficial result (increased pulsatile progesterone production) that helps to correct for another aspect of reduced granulosa cell function (lower tonic progesterone production). In this feedback loop we can see a compensatory mechanism that may serve to buffer female fecundity from certain aspects of declining function in the aging ovary.

Nevertheless, it is clear that female fecundity does decline, at least on average, with increasing age through the thirties and forties, as the cumulative evidence from ART procedures makes clear. Particularly intriguing are the results of success rates in programs of in vitro fertilization with donor eggs. One difficulty with the interpretation of in vitro fertilization studies in general is that hormonal profiles are artificially manipulated; thus the effects of age on one of the two primary functions of the ovary, hormone production, are removed from consideration. As noted in the discussion of the report by Mark Sauer and his colleagues, the primary conclusion to be drawn from these studies is that uterine function in older women is not limiting in an ultimate sense. One study, however, by David Levran and his colleagues in Israel attempted to assess the degree to which the age of both donor and recipient independently affected the results of donor egg IVF trials. They found that the age of the donor strongly affected the probability of successful fertilization and implantation, while the age of the recipient significantly affected the rate at which pregnancies, once established, were lost. It is tempting, then, to associate the age of the donor with "egg factors" and the age of the recipient with "uterine factors." In the latter case, however, we cannot discount the possible contribution of chronically low ovarian steroid stimulation over the period prior to the clinical intervention.

Hormonal production is not the only aspect of ovarian function that shows changes with advancing age in mid-life. Gamete production is affected as well. Ovulatory frequency declines perceptibly after age thirty-five and notably after age forty. There is also evidence of declining quality among the oocytes that are produced. Rates of genetic disorders and congenital malformations increase with advancing maternal age. Particularly dramatic is the increase in cases of chromosomal nondisjunction. This occurs when the two copies of a chromosome are not separated during meiosis, with the result that two copies, or none, get into the egg that is produced. When this egg is fertilized, the resulting embryo has one or three copies of the affected chromosome rather than the normal two. Serious developmental problems can arise from this incorrect chromosome number, many of which are fatal and probably contribute to the increasing rate of early embryonic loss with increasing maternal age. Other cases of chromosomal nondisjunction may result in viable offspring, although the affected individual may suffer from numerous anatomical and developmental abnormalities. Trisomy 21, or Down syndrome, is one of the most common and most widely known of these chromosomal disorders. Rates of trisomy 21 increase exponentially with maternal age, from less than 1/3,000 at age twenty-five to 1/365 at age thirty-five to 1/32 at age forty-five. In many countries screening for trisomy 21 and certain other disorders is now routine for pregnant women over thirty-five. It should be noted that there is evidence of increasing rates of chromosomal nondisjunction in sperm with advancing male age as well. However, because there are so many more sperm competing for the chance to fertilize the egg, the effect of male age on chromosomal anomalies in the embryo is much weaker than the effect of female age.

Thus a picture of steadily declining ovarian function during the last decade of reproductive life should be added to the picture of steadily increasing ovarian function during the first decade. Yet these patterns are not simple mirror images of each other. Ovarian function increases more steeply in the first decade than it declines in the last. Further, while all aspects of ovarian function in young women seem lower relative to mid-aged women, the ovarian function of older women shows evidence of compensatory mechanisms that serve to sustain female fecundity despite the possible negative consequences of a dwindling pool of aging gametes and associated granulosa cells.

But, most important, we must recognize the limitations imposed by the cross-sectional nature of all the available data on age-related fecundity decline. Virtually all our data represent age-group averages and trends, not the trajectories of individual women. It is possible that a cross-sectional pattern of gradual decline in ovarian function with increasing age could be produced from superimposing individual, longitudinal trajectories, each of which has a more abrupt threshold of rapid decline, but which differ in the age at which the threshold is crossed. These two alternative possibilities concerning the way in which both individual ovarian function and individual fecundity change with age, smoothly or abruptly, cannot easily be discriminated on the basis of cross-sectional data.

There are some available data that favor a gradual decline in ovarian function before menopause rather than an abrupt decline at the end. This information was obtained by using a measure of FSH from the early follicular phase to discriminate women who are on the verge of the menopausal transition. A higher than average level FSH in the first days of a menstrual cycle seems to typify women whose ovaries are failing to produce enough estradiol and inhibin to suppress pituitary gonadotropins. This condition is usually considered indicative of an "insufficient follicular reserve," meaning that the woman is beginning to run out of follicles. Alternatively it may represent a further progression of the decline in ovarian function that leads first to increases in LH and pulsatile progesterone secretion. In either case it is evidence of impending menopause. Women who show this elevated FSH pattern have particularly poor chances of becoming pregnant through IVF, and most ART clinics screen older patients for this diagnostic indicator. The same indicator can be used, however, to remove peri-menopausal women from an analysis of the effect of age on IVF success rate. When this is done, the effect of age remains highly significant and just about as steep as before. This finding suggests that female fecundity does decline with age in a progressive way, and that the gradual decline in fecundity in cross-sectional studies is not just an artifact of the staggered entry into an abrupt menopausal transition by individual women.

Assuming there is a gradual decline in ovarian function and individual fecundity in women during the decade or so preceding menopause we can ask, as we did about the increase in ovarian function in adolescent women, whether there is any functional, adaptive significance to the pattern. As we have seen, the standard prediction of life history theory is that reproductive effort should increase with age. And yet, at the end of the female reproductive span, we find natural birth spacing increasing and evidence of declining ovarian function. The decline in ovarian function might be a physiological consequence of aging gametes and granulosa cells, as noted above. In this case the organism may be paying a cost late in life for an advantage gained at an earlier period. An alternative view would suggest that the life history trade-off that is faced during the later part of a woman's reproductive career is not primarily a choice between investing in her own survival and producing offspring, but increasingly a choice between investing in new offspring and investing in offspring already born. The older a woman gets, the argument would go, the more her fitness can be enhanced by ensuring the survival and reproduction of offspring already born rather than attempting to bear and raise yet another child. This argument would become even stronger if the risks of successful reproduction decreased in some intrinsic way with maternal age, owing, for example, to declining gamete quality or deteriorating maternal physiological vigor. In this regard it is notable that birth weights decline and the risks of fetal and neonatal death increase with increasing maternal age even among the most privileged populations.

Whether declining female fecundity with advancing age can be viewed as adaptive would remain a significant question even if menopause did not occur, bringing fecundity abruptly to zero. This more dramatic change in fecundity, however, often preempts any theoretical consideration of age and female fecundity, in part because the apparent evolutionary paradox seems so stark.

RUNNING DOWN OR RUNNING OUT?

Like all biological phenomena, the final cessation of ovarian function in older women requires both a proximate (mechanical) and an ultimate (functional) explanation. In a proximate sense, menopause is the result of a depletion of the ovarian supply of follicles and eggs. This depletion was first demonstrated in humans by Erik Block through the histological examination of the ovaries of accidental death victims, and has been confirmed more recently in studies of ovarian histology after voluntary ovariectomy. With no follicles to respond by producing estradiol, pituitary gonadotropin production is unrestrained by negative feedback, resulting in high amplitude fluctuations of both FSH and LH in menopausal women. These gonadotropin surges have been linked to the "hot flashes" that many women experience during the menopausal transition, and that can be documented by accurate skin and core body temperature recordings. The fact that this pattern of pituitary activity can be corrected by the administration of exogenous steroids and the fact that successful pregnancy can be achieved with donor eggs clearly implicate failure of ovarian function, rather than changes in hypothalamic or pituitary function, as the proximate cause of menopause.

Ovarian depletion in turn is a consequence of the conjunction of two other traits: a fixed oocyte supply and a long life-span. A woman's lifetime gamete supply is established in the second trimester of her own fetal development, when the germ cell line ceases mitotic proliferation. Each of the several million primitive germ cells, called oogonia, that exist at that point is enveloped in a layer of granulosa cells, begins the first stages of the first meiotic division, and then goes into developmental arrest. From that point on the follicular supply is never increased, but rather begins a process of steady, exponential decay. Every month some of the remaining follicles leave the "resting pool," as is evidenced by increasing metabolic activity on the part of the granulosa cells, and start on a developmental path that leads either to ovulation or to atresia (cell death). The process of leaving the resting pool itself appears to be independent of the activity of the hypothalamic-pituitary-ovarian (HPO) axis, since it occurs continuously, regardless of maturational or reproductive state, during childhood, pregnancy, and lactation, as well as during periods of cyclic ovarian activity. At reproductive maturity a woman retains only some 75 percent of the gamete supply she had at birth (which is itself only 33 percent of the maximal supply attained in the second trimester of gestation). By age forty she has less than 2 percent of that maximal supply. Recent evidence suggests that the rate of ovarian depletion may accelerate in the final peri-menopausal period, perhaps reflecting the increase in FSH levels characteristic of a declining ovarian reserve. By this point the ovarian follicular supply may be too low to support regular cycles. If, as seems likely, the processes of recruitment that culminate in the selection of the dominant follicle in some degree select for gamete quality, then the exponentially increasing frequency of poor-quality ova produced in late reproductive life may in part be a consequence of the decreasing pool of gametes to select from.

As a contrast to the human pattern, consider the cause of reproductive failure in laboratory mice. Females of several strains regularly undergo a cessation of reproductive cycling if they are kept alive long enough, but in their case the cause seems to be hypothalamic, rather than ovarian. Continued exposure to periodically high estrogen levels appears to result in a progressive desensitization of the mouse's hypothalamus, so that ever higher levels of ovarian steroids are necessary to effectively elicit a positive feedback response. For the wild mouse, life is ordinarily so short that follicular supply is never limiting. But when life is artificially prolonged under laboratory conditions, this progressive process results in a hypothalamus so insensitive to estrogen feedback that a positive feedback surge of gonadotropins to trigger ovulation is no longer possible. The proof of the fact that the ovary is not responsible is provided by transplant experiments where the ovaries of young mice are transplanted into the bodies of old mice, and vice versa. Young ovaries cease to cycle when coupled with an old hypothalamic-pituitary axis, whereas old ovaries resume cycling when coupled to a young hypothalamic-pituitary axis.

Human menopause would presumably not occur, then, if the female gamete supply were continuously or periodically replenished, or if women had appreciably shorter life-spans. Analogies to these hypothetical conditions are easily identified. Men provide an example of the first case, a continuously replenished gamete supply, since male germ cells continue mitotic proliferation throughout a man's life. As a consequence, although male reproductive capacity in general, and testicular function in particular, decline with advancing age, they do so along a trajectory similar to the pattern of general physical senescence, not at all like the pattern of early and independent cessation shown by female ovarian function.

Many other mammals can be cited in support of the second possibility. Trajectories of ovarian depletion have been documented for rats, cows, and rhesus monkeys, all predicting follicular exhaustion at ages close to observed maximal life-spans of wild populations of the same or related species. As we shall see presently, presumably any mammal *can* undergo menopause if it lives long enough, since all have finite follicular supplies. Yet menopause in the wild may be relatively rare owing to the close congruity of the reproductive span and the life-span in most species.

If, in a proximate sense, human menopause is a consequence of a finite oocyte supply in conjunction with a long life-span, then it is likely that the ultimate explanation lies in these areas as well. Indeed, some have argued that menopause is a recent phenomenon in human evolutionary history, that prehistoric humans, like most other mammals, would rarely have lived much beyond the point of ovarian depletion. It is likely that appreciable survivorship beyond the age of menopause is more ancient than those making this argument suppose. Nevertheless, it is almost certainly true that a relative extension of life-span occurred sometime between the hominid divergence from the other African apes (some six to eight million years ago) and the present. Yet in this fact lies a formidable evolutionary problem. How could evolution produce an extension of life-span without producing an extension of reproduction?

The pronounced divergence of human life and reproductive spans directly challenges the prevailing evolutionary theory of senescence. According to this theory, which has a precise quantitative expression, the power of natural selection to prolong life (or to eliminate genes that contribute to increasing mortality risk at older ages) is a function of reproductive value, or the expectation of future offspring: only if there is some difference in reproductive success between an individual who dies older and an individual who dies younger will natural selection be able to operate. If reproduction ceases at fifty, what fitness differential can exist between a woman who dies at sixty and one who dies at seventy?

This theoretical conundrum is often referred to as "the problem of the evolution of menopause," but it ought properly to be called "the problem of the evolution of postreproductive life." If one imagines that an extended life-span evolved for separate reasons (as a correlate of in-

creased brain size, for example), one still has to account for the failure of the reproductive span to adjust. Presumably there would be a significant advantage in fitness to any variation that would increase the reproductive life-span of a female under such conditions. Some argue that the increase in life-span has been too recent for natural selection to have affected the age at menopause, and that voluntary limitation of fertility now makes such selection unlikely. These arguments run afoul of evidence that suggests a substantial time depth to the modern human life-span. More popular have been arguments that postreproductive females *do* in fact continue to contribute to their own fitness, through continuing investment in their own young children as well as through investment in their grandchildren and other relatives. A woman who dies at sixty is likely to have fewer descendants than a woman who dies at seventy, according to this way of thinking, because her own immature children are less likely to survive and her mature children are less likely to reproduce successfully without her. Kristen Hawkes and her colleagues have provided impressive evidence, for example, of the substantial contributions that postmenopausal women make to provisioning their offspring and grand-offspring among the Hadza, a hunting and gathering group in Tanzania.

It is easy to believe that postreproductive women do continue to contribute to their own fitness in these ways, but it is harder to believe that they contribute *enough,* in a quantitative sense, to outweigh the advantages of extended reproduction. Kim Hill and Magdalena Hurtado used demographic data from the Ache, a group of South American foragers, to test the logic of the so-called grandmother hypothesis in quantitative terms. They concluded that it is very unlikely that such indirect fitness contributions could ever be strong enough to offset the selection pressure for extended reproductive life.

If menopause cannot be fully understood as a byproduct of the evolution of a longer life-span, perhaps it can be understood in terms of the functional significance of a finite oocyte supply. In fact, the pattern of oocyte production that humans display seems to demand explanation in many ways. Why don't females support continuous gamete proliferation in the male pattern? If there is going to be a finite supply of gametes, why establish it so early in development? If the supply is to be established so early, why draw on it continuously beginning more than a decade in advance of reproductive maturation and continuing through extended nonfecund periods?

Some insights can be gained from considering the phylogenetic distribution of restricted oogenesis (the production of oocytes) and recurring oogenesis. Within the vertebrates, recurring oogenesis is clearly the more primitive condition, occurring in most fish, amphibians, and reptiles, with restricted oogenesis characterizing birds and mammals. One interesting exception to this rule is the subgroup of cartilaginous fish, the sharks and rays, which do have restricted oogenesis. These fish are also conspicuous as a group for having highly developed internal gestation, in some cases even involving the development of a true placenta. Rather than spewing hundreds of eggs at a time into the sea, they give birth to a small number of live young. The implication is that restricted oogenesis ordinarily exists in association with a pattern of heavy maternal investment in smaller numbers of offspring.

It makes sense that an animal that depends on producing large numbers of embryos, each with a small survival probability—a barnacle for example—would profit from an ability to continuously renew her gamete supply. It is less obvious why an animal that produced fewer embryos of higher survival probability would necessarily profit from restricting her gamete supply. Some energy savings might be realized from curtailing excessive proliferation of gametes, assuming that the production of extra gametes is "excessive production." If some gamete selection occurs during the process of follicular development, then "overproduction" of gametes may be necessary to promote embryo quality. On the other hand, truly excessive gamete production could lead to higher rates of mutation and ultimately lower embryo quality since chromosome replication errors accumulate in proportion to the number of cell generations. These countervailing advantages and disadvantages of overproduction of oocytes may result in an optimal number of gametes for attaining the highest possible embryo quality.

Similar logic may apply to the rate of follicular atresia. Slowing the rate of atresia would allow a given supply of gametes to last longer and hence to postpone menopause. The rate of atresia is primarily determined by the rate at which follicles leave the resting pool to resume metabolic activity. This rate determines the degree of gamete selection that can be realized in any given ovarian cycle. Slowing the rate of atresia might therefore entail a cost in terms of embryo quality. An embryo that results from the best egg out of a dozen may not be the equal of one resulting from the best egg out of hundreds. A dwindling follicular pool at the end of reproductive life may contribute in this way to the higher rate of spontaneous abortion and chromosomally anomalous embryos.

But even if restricted oogenesis makes sense in terms of energy savings and the preservation of gamete quality, why push the process so early in development? It appears to be generally true of both birds and mammals that oocyte numbers are established before birth or hatching, even in species whose reproductive maturation is years off. The answer here may be that the fetus in the womb or in the bird egg is more protected from mutagenic influences in the environment than it will be after birth or hatching. Hence the gamete quality of individuals who complete oocyte proliferation during fetal life may be higher than that of individuals who continue or delay this process until later.

Most difficult seems to be the question of continuous depletion of the oocyte supply once it is established. Surely further economies could be achieved by limiting oocyte recruitment to the appropriate fecund periods. Perhaps the cost of the machinery necessary to achieve this degree of facultative recruitment is simply greater than the resulting energy savings and boost in embryo quality could justify. The length of time that passes between the initial emergence of a follicle from the resting pool and its potential ovulation appears to be as great as three to six months. Thus the mechanisms necessary to achieve facultative recruitment would need to be able to anticipate the state of the organism at least that far in advance.

Although many questions regarding the mechanisms of oogenesis remain to be answered, a general hypothesis can be formed as follows. Restricted oocyte production evolved as a correlate of increased maternal investment in the fetus, with particular benefits in terms of the preservation of gamete quality. In natural populations of birds and mammals restricted oogenesis does not exact a heavy selective cost, because the natural life-span rarely extends beyond the point of ovarian depletion. Early hominids inherited this pattern of gamete production as part of their phylogenetic heritage. When the human, or proto-human, life-span became significantly extended (for reasons that may be entirely separate from our present concerns), menopause became a phenotypic (outward) expression of ovarian depletion. To this extent menopause is a temporally delayed correlate of a separate trait that occurs much earlier in the life of the organism, the fetal establishment of a finite oocyte supply. Although menopause itself may carry a selective cost, the earlier trait with which it is linked may carry a selective benefit. Such temporally linked costs and benefits are an example of the "antagonistic pleiotropy" on which George Williams originally based his evolutionary theory of senescence. Traits with early benefits can evolve despite later costs, Williams argues, because of the necessary decline in the force of selection resulting from intervening mortality and reduced opportunity for further reproduction. Thus mammals and birds may have evolved restricted oogenesis for the benefits that accrue early in life despite the small risk of running out of eggs before they die.

Thus we don't need the grandmother hypothesis to explain why menopause "evolved" in humans. The trait with which menopause is pleiotropically linked, restricted oocyte production, evolved many millions of years before humans. Rather, what demands explanation in humans is the *maintenance* of menopause in the face of a long period of postreproductive life. Here the grandmother hypothesis may be more helpful. Delaying ovarian depletion may after all not be an "easy" thing to accomplish. Owing to the steep exponential decay in the resting pool, rather large increases in the initial oocyte number would be necessary to achieve relatively small temporal extensions of an effective oocyte supply. Dramatic increases in oocyte supply would probably necessitate a much larger ovary, disproportionate to body size, and larger fimbria on the oviducts to collect eggs ovulated from a larger ovarian surface. All of these adjustments would presumably carry energetic costs. Alternatively, the rate of oocyte recruitment could be slowed; however, this would probably entail some sacrifice of gamete quality because it would reduce the potential for gamete selection, as discussed above. Even if we assume that a net benefit from such adjustments would still accrue, it might accrue at a relatively slow marginal rate. If the advantages of "grandmothering" accrue at a faster marginal rate than the advantages of redesigning the effective oocyte supply, they could produce a "local adaptive peak," in the terms of Sewall Wright: Even if the ultimate advantage that could be accrued through redesigning the oocyte supply is greater than the current advantage, it may not be possible to get there from here without passing though intermediate stages of diminished fitness. The social organization of hominids may have provided for such a local adaptive peak by providing the opportunity for postreproductive females to continue to contribute to their own fitness through favoring the survival and reproduction of their kin. In this regard, it is interesting that the only other mammals with a well-documented evidence of a high frequency of natural menopause and extended postreproductive life are toothed whales, animals whose intelligence, complex social organization, and long-term kinship relationships may provide similar opportunities. In contrast, the baleen whales, such as the fin whale, which are much more solitary, appear to have taken the other route. Long life-spans in baleen whales are associated with large body sizes, large ovary sizes, and long reproductive life-spans.

This way of thinking essentially turns the grandmother hypothesis on its head. Given the evolution of postreproductive life, there would be tremendous selective pressure on females to behave in ways that would contribute to their fitness through nonreproductive channels. Postreproductive life, in this view, selects for grandmaternal investment in kin, rather than the other way around. But what could have selected for the increase in life-span in the first place? And why did the increase in life-span not involve a simultaneous increase in reproductive span? I think the answer to the second question is that the extension of life-span was accomplished without a significant change in body size. Normally life-span and body size scale together within a given evolutionary lineage, and oocyte supply may simply scale with body size. The extension of human life-span seems to come from some other quarter as yet unidentified. It is tempting to speculate that some evolving aspect of complex social organization led to the change by decreasing the vulnerability of group-living hominids to extrinsic sources of mortality such as predation, starvation, or disease. Comparison of mortality rates between contemporary foraging populations of humans and wild populations of chimpanzees suggest that compared to chimpanzees and other apes we are particularly good at surviving into adulthood. The keys to the evolutionary puzzles of menopause and old age may thus lie in the survival risks faced by infants and children.

ALL TOGETHER NOW

The age trajectory of ovarian function and female fecundity appears to be the product of different selective forces acting at different phases of life. Some of these selective forces, such as those resulting in restricted oogenesis, are extremely ancient. Others, such as those associated with the extension of life-span beyond ovarian depletion, are comparatively recent. What unfolds as whole cloth in the reproductive careers of individual women has been stitched together by diverse causes. Two reasonable questions follow from this realization. Do the different stages of female reproductive life vary independently across women, or is there an inherent integrity to the reproductive course such that ovarian function is correlated at different ages? And does the age pattern of ovarian function unfold similarly in different populations and different environments, or are different circumstances associated with different patterns of ovarian aging?

Some evidence, alluded to at the end of the last chapter, is beginning to accumulate to suggest that there is a degree of integrity to ovarian function across the reproductive span, although the total amount of data on this point is still quite limited. One source of evidence consists of the longitudinal studies of ovarian function in adolescence carried out by Apter, Vihko, and their colleagues. Motivated by a desire to understand the epidemiological association between early menarche and increased risk of breast cancer, these researchers followed the development of ovarian function in a group of 200 Finnish girls. Their results suggest that girls with late ages at menarche have absolutely slower rates of increase in ovulatory frequency than their early maturing peers for many years beyond menarche. While girls who reached menarche before age twelve were ovulating 80 percent of the time within two years, girls who reached menarche after age thirteen were still ovulating only 65 percent of the time six years after menarche. Follow-up data at later ages suggest that the later maturing women may continue to have lower levels of ovarian function as a group than their early maturing peers do, well into the peak reproductive years. In another longitudinal study of adolescent ovarian function, Stefano Venturolli and his colleagues found that late maturers are more likely to show irregular and abnormal hormonal profiles later in life than are early maturers. In a retrospective study, J. Gardiner and Isabel Valadian found that late maturers were more likely to experience oligomenorrhea and dysmenorrhea throughout their reproductive careers than were early maturers.

To these studies of variation in ovarian function over the reproductive span within populations may be added studies of variation between populations. A great deal of attention has been paid to variation in menarcheal age between populations. An association between earlier maturation and higher socioeconomic status within populations, and a secular trend toward earlier maturation being a common correlate of economic development, paralleling the demographic transition to lower mortality, are common observations. The relationship between these changes in age at menarche and adult fecundity is not well studied, however. In general, the demographic transition to lower levels of mortality over time tends to be accompanied by a transition to lower, controlled fertility. As often noted before, however, fertility patterns under conditions of widespread and effective contraceptive practice are not expected to reflect patterns of fecundity. For those patterns we have to look to natural fertility populations.

The data that have been collected on the Lese of the Ituri Forest, the Tamang of central Nepal, rural peasant women from Poland, Quechua women in highland Bolivia, and middle-class women in Boston provide a limited opportunity to address some of these issues. A comparison of the reported average age at menarche for these populations with levels of luteal progesterone among adult women aged twenty-five to thirty-five reveals a rather tight negative correlation. Women who represent early maturing populations, like those in Boston, have higher levels of ovarian steroid production as adults than do women representing later maturing populations, like the Lese and the Tamang. Of course, acute conditions in these different populations may contribute to these differences among adults as well as the persistent effects of developmental differences. But the extreme parallelism between the changes in the indices of ovarian function with age for such distinct populations as the Tamang, the Lese, and Boston women suggests a developmental integrity to ovarian function throughout life.

Variation in age at menopause between populations is much more difficult to study than variation in age at menarche. For one thing, menopause is not as susceptible to concrete definition as a discrete event. Typical definitions in clinical and epidemiological studies refer to the absence of menses for six months or a year before menopause is officially recognized in a given woman. Even then, a single menses can cancel out the previous year of amenorrhea. Certainly some women experience episodes of amenorrhea that long or longer in midlife for reasons other than menopause, episodes that are terminated by the resumption of menses. Finally, in many natural fertility societies lactation following the birth of a final child may easily last several years and menses may never resume. When menopause occurred in such an interval of extended amenorrhea is anybody's guess. Hence the data on population variation in menopause are of questionable reliability at best.

Our understanding of age variation in ovarian function in fact suggests that the timing of follicular depletion may well be independent of other features of the arc of reproductive life. Since follicular depletion is a function of the initial supply and rate of attrition of follicles, the only way the environment or a woman's own reproductive history could affect the timing of this event would be by affecting one of those two components. Reproductive state and reproductive history appear to have no effect on the rate of follicular attrition, and necessarily can have no effect on the size of the initial follicular supply, which was established a decade or more before the beginning of the reproductive career. Environmentally induced variation in age at menopause is still theoretically possible, but it would need to be shown that either the initial follicular supply or the rate of follicular attrition varied in some predictable way with the environment. Follicular attrition rate seems particularly buffered from such perturbations, and little is known, for obvious reasons, about variation in the initial follicular supply and its potential environmental correlates.

Age at menopause, however, may vary independently of age at follicular depletion to a limited degree. Strictly speaking, menopause is the final cessation of menstrual bleeding. Because endometrial proliferation is stimulated by follicular estradiol production, depletion of follicles normally represents a final limit on natural menstruation. Menstruation may, however, cease before the final limit of follicular depletion is reached if follicular estrogen production falls below a threshold sufficient to support endometrial growth. To this extent, women on a lifelong lower trajectory of ovarian function by age may experience their final menstrual period before women on a higher trajectory, even if both groups of women actually reach the point of follicular depletion at the same age. As noted above, we sorely lack longitudinal data on late reproductive life with which to test this possibility. It would, however, be consistent with the rest of our understanding of ovarian function through life.

If there is a developmental integrity to ovarian function over the arc of life, we can question whether it has any functional significance, or is simply a manifestation of nonfunctional human variation. One possibility suggested at the end of the last chapter, merits repeating here. Natural selection may have designed developmental mechanisms to adjust the average level of female fecundity to chronic conditions of energy availability. This would follow if the optimal rate of reproduction in energy-poor environments were lower than the optimal rate of reproduction in energy-rich environments. Using the period of growth and development as a "bioassay" of chronic environmental conditions and establishing a lifetime set-point for baseline ovarian function on that basis is one way of achieving this result. If the environment can only support slow growth and maturation, it may only be able to support a slow average reproductive rate as well. Variation in ovarian function about the baseline in response to changes in environmental conditions would still be possible.

Clearly, a great deal more information is needed before the question of the integrity of ovarian function across the reproductive span can be satisfactorily answered. At this point we can only say that it is a reasonable hypothesis. But it is an important hypothesis to confirm or refute. Lifetime exposure to ovarian steroids has a very strong impact on the risk of reproductive cancers in women, including breast cancer. The dramatic reduction of estrogen at menopause is associated with accelerated bone loss and increased risk of heart disease. All of these serious contemporary health risks may be strongly affected by the trajectories of lifetime ovarian function that women follow, and our understanding of those risks and the opportunities for intervention may depend on our understanding of the trajectories of ovarian function and how they are determined. Much of what we need to know will be difficult and costly: longitudinal information on ovarian function within individual women, comparable data on ovarian function from a broad spectrum of human populations, information on variation in the size of the initial follicular supply in individuals and in the rate of follicular attrition. Paradoxically, it seems we are gaining the ability to artificially control the very shape and limits of our reproductive lives before we fully understand what it is we are controlling.

THE BODY BUILDERS

TO SAY THAT MEN AND WOMEN differ in many details of their reproductive physiology is obvious and boring. So let's start with the opposite statement: men and women are remarkably similar in many details of their reproductive physiology.

The central hormonal axis governing gonadal function is the same in both sexes. The hypothalamus releases GnRH into the hypophyseal portal system in quasi-hourly pulses in males just as it does in females. Alteration of the pulsatile GnRH pattern in the direction of either too many pulses or too few can result in a disruption of gonadotropin release by the pituitary gland. Pubertal development normally involves the appearance of the mature pulsatile pattern of GnRH. Exogenous manipulations of GnRH pulsatility—either introducing GnRH pulses where they are absent, or suppressing them (via long-acting GnRH analogs) when they are present—can advance or retard pubertal progression in individuals with developmental pathologies of hypothalamic origin.

When GnRH pulses occur in the mature hourly pattern, the male's pituitary responds, as does the female's, by releasing gonadotropins. These gonadotropins are the same ones found in females—follicle stimulating hormone (FSH) and luteinizing hormone (LH)—even though males have no follicles to stimulate or to luteinize. In early physiological studies, these hormones were given different names in males, referring to their principal functions in the male reproductive system—interstitial cell stimulating hormone instead of follicle stimulating hormone, and Leydig cell stimulating hormone instead of luteinizing hormone.

When the composition of these large and complex protein molecules was finally worked out, however, the male and female gonadotropins were found to be identical, and the female names were given precedence.

Not only are the structures of the gonadotropin molecules the same in males and females, but their targets and actions are analogous as well. To appreciate this similarity we must first appreciate the anatomical and histological analogies between the ovary and the testis. The gonads develop early in embryogenesis around a duct system (the Wolffian ducts) that leads into the developing urinary tract. A second duct system (the Müllerian ducts), outside the developing gonad, also leads to the urinary sinus from the open body cavity near the gonad. The precursor cells to the gametes migrate to the developing gonad from their site of production near the developing umbilical cord. As they reach the gonad, which to this point is identical in males and females, the first expression of genes for sex determination comes into play. The proliferating gametogenic cells either locate themselves in the outer layer, or cortex, of the gonad, or migrate deep into the interior, or medulla, of the gonad to line the system of tubal ducts. This positioning of the gametes determines the direction in which they will travel later in the organism's life in their quest for gametes of the opposite sex. Those gametes that are located in the cortex will be shed to the exterior of the gonad to be transported by the fallopian tubes (the erstwhile Müllerian ducts) to the uterus. Those gametes that are located in the medulla of the gonad will be shed internally into the semeniferous tubules, thence to the rete testis, epidydimis, and vas deferens (derived from the Wolffian ducts).

The location of the gametogenic cells is the first anatomical distinction between ovary and testis. Subsequently the unused duct system in each sex degenerates to provide another distinction. In each location, however, the gametogenic cells become separated from other tissues and from the body's blood supply by a surrounding membrane that seals off tissues on either side. This kind of membrane is called a basement membrane. Such a membrane defines the follicles in the developing ovary and the seminiferous tubules in the developing testis. Inside the membrane together with the developing gametes are specialized "nurse" cells—granulosa cells in the female, Sertoli cells (also called interstitial cells) in the male. These nurse cells maintain cytoplasmic connections with the gametes as they develop through which important nutrients and signals for development flow. Outside the membrane in each case is a second population of cells that is responsible for the production of steroid hormones necessary for normal gamete maturation. These are theca cells in the female and Leydig cells in the male.

In the female, the theca cells, unlike the granulosa cells, are able to make steroid hormones from scratch, that is to say, from cholesterol. The theca cells respond to LH pulses from the pituitary by producing androgens, primarily testosterone. So it is with the Leydig cells in the male, which make testosterone de novo from cholesterol in response to LH. In the female, the granulosa cells inside the basement membrane support the development of the egg cells and convert testosterone derived from the theca cells into estradiol. So it is with the Sertoli cells in the male, which nurse the developing sperm cells and convert testosterone to estradiol. When follicles are growing and gamete maturation is under way, the granulosa cells in the female also release a protein hormone called inhibin, which, upon reaching the pituitary, selectively suppresses the release of FSH. So it is with the Sertoli cells in the male. As long as spermatogenesis is under way, the Sertoli cells produce inhibin, which feeds back negatively on FSH release to maintain very low circulating levels. In the female, FSH is necessary to initiate follicular maturation at the start of each menstrual cycle. But when steroid production by the dominant follicle reaches a sufficient level, high levels of FSH stimulation are not required for follicular growth and gamete maturation to proceed. So it is in the male. FSH is required to initiate gametogenesis, or to restart it if it becomes interrupted. But once established, steroid production is sufficient to keep the process going with only minimal FSH. If spermatogenesis is interrupted, by starvation for example, inhibin production is interrupted as well. When conditions improve and GnRH pulses begin again, the pituitary releases FSH in

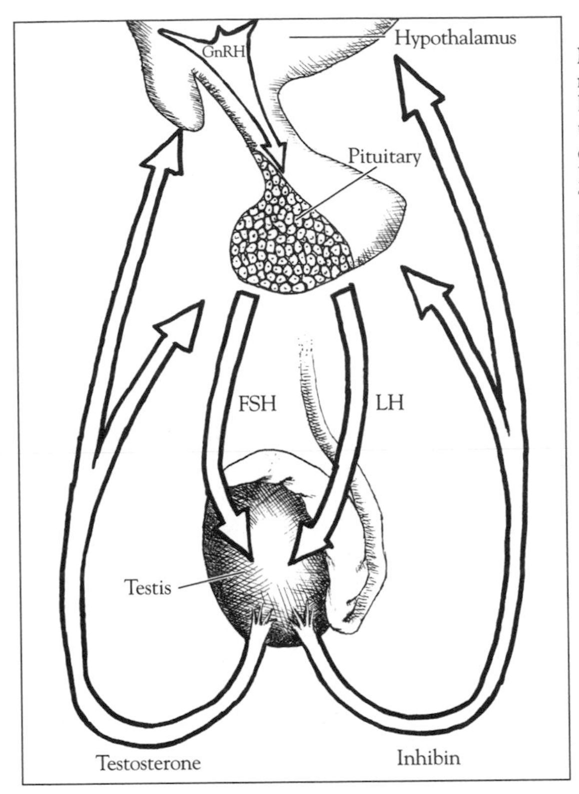

Major components of the male reproductive axis. GnRH is released in pulses from the hypothalamus, stimulating the production and release of FSH and LH from the anterior pituitary. Together FSH and LH stimulate production of sperm and production of the hormones testosterone and inhibin. Testosterone modulates hypothalamic and pituitary activity while inhibin suppresses the production of FSH.

quantity as well as LH until spermatogenesis is reestablished. After that, only a low circulating amount of FSH is necessary to sustain aromatization of testosterone to estradiol by the granulosa cells or Sertoli cells in either sex.

This recitation of the similarities of male and female reproductive physiology helps to set the important differences in relief. Among those differences two are paramount. Males produce gametes in prodigious quantities, not one at a time, and do so continuously, not at monthly intervals. All along the length of the semeniferous tubules (which would extend for dozens of meters if uncoiled) mitotically active spermatogonia divide to produce sperm cells. The developing sperm cells undergo an extended process of maturation under the nurturance and guidance of the Sertoli cells, migrating in the process from their site of origin next to the basement membrane toward the open lumen at the center of the tubule. The entire course of maturation takes around seventy-four days, but it proceeds in asynchronous waves along the length of the seminiferous tubule, and new crops of sperm cells are mitotically produced in any given location before the previous crop has completed its course. The result is a continuous flow of mature sperm cells into the lumen of the semeniferous tubules, from which they are collected into the epidydimis to await ejaculation or eventual death and resorption.

Superficially, the maturation of a sperm cell looks like the opposite of the maturation of an egg cell. Where the egg cell increases in size and accumulates cytoplasm, the sperm cell shrinks in size and sheds cytoplasm. It assumes a highly specialized morphology dedicated to the task of fertilization. In the end, it consists of a head containing the paternal set of chromosomes topped by a special cap of enzymes known as the acrosome. The enzymes of the acrosome will be used to digest the veils of cellular and secreted material that surround the egg, including the cumulus oöphorus and the zona pellucida. Behind the head is the midpiece of the sperm cell in which are packed all the cell's mitochondria, the engine that will provide energy to the propeller. The propeller is represented by the tail of the sperm cell, a long, mobile filament comparable in form and function to the flagellae that many unicellular organisms use to propel themselves through their aqueous world.

Stripped to the bare essentials, the sperm cell depends on its immediate environment to provide it with nearly everything it will need to complete its mission of fertilization. The seminal fluid, which is produced by the prostate gland and seminal vesicles, is rich in simple carbohydrates that the mitochondria of the midpiece can use for fuel, as well as carbonates to neutralize the acidic environment of the vagina. Once through the cervix into the uterine cavity, the sperm cell will rely on nutrients provided by the secretory glands of the endometrium for fuel. Although the sperm cell has a propeller of sorts, it has no rudder. It depends on its immediate environment to provide it with direction as well as fuel. In the middle of the menstrual cycle, under the influence of LH, the mucus that normally fills the cervix becomes thin, and the individual mucin strands become stretched to lie parallel to the axis through the cervix, rather than presenting a tangled mass as they do during most of the rest of the menstrual cycle. The sperm cells use the parallel tracks provided by the mucin strands to orient their swimming. The thermal oscillations produced in the mucin strands by the woman's body heat also resonate with the beating of the sperm cell's tail to aid its progress. Further up the woman's reproductive tract the steady current of fluid passing down the fallopian tube toward the uterus—a current produced by the beating of the microscopic cilia lining the tube and carrying the egg cell toward its site of potential implantation—works like wind on a weather vane to keep the sperm cells swimming upstream.

A LIMITING DIFFERENCE

It's a hazardous journey, and only a small fraction of the sperm cells that enter the vagina at ejaculation ever arrive in the vicinity of the oocyte to have a chance at fertilization. This fact alone might seem sufficient to account for the high rate of production of male sperm cells. But it really is only part of the story. It may help to remember that continuous production of a large number of gametes is the primitive pattern inherited by all vertebrates from their invertebrate ancestors. It is a pattern that female mammals have abandoned for specific reasons tied ultimately to their necessarily heavy physiological investment in relatively few offspring. The fact that males still follow the ancient pattern of gamete production suggests that male reproductive success is not subject to the same energetic constraints. Because male mammals do not gestate or suckle their own offspring, male reproductive output is not limited by the rate at which nutrients can flow across the placenta or through the nipple.

What primarily limits male reproductive output is not the rate at which energy can be converted into offspring, but the number of opportunities to mate with fecund females. This may seem a rather crass statement, but it is a generalization that is fundamental to male reproductive physiology. Among the consequences that follow from this basic difference in constraints on male and female reproductive success are three that shape male fecundity in patterns distinctly different from those of female fecundity.

The first consequence is the fact that males produce a prodigious and nearly continuous supply of gametes. Female mammals cannot exceed a reproductive rate of one litter at a time, even theoretically. Males, however, can sire numerous overlapping litters, if opportunity allows. In order to realize such opportunities, gametes must always be available. Only when it can safely be assumed that all females are infecund, as in the nonbreeding seasons of some species, is sperm production suspended in healthy adult males. Even when the chance of a successful mating is low, the cost of missing that opportunity when it comes along is likely to be greater than the cost of maintaining the necessary sperm supply. Even in a species like humans that breeds year-round, an individual female's reproductive lifetime is normally composed of alternating fecund and infecund intervals. Fecundity in human males, by contrast, is normally continuous and uninterrupted.

A second consequence is related to the first. Male gamete production shows little variation with age other than that associated with general physiological senescence. In humans, gamete production begins at puberty and reaches adult levels within a few years. It is sustained at that level in healthy individuals from early adolescence until late middle age, when it begins to decline in parallel with other physiological systems. Unlike female fecundity, which increases gradually from menarche until the mid-twenties and then begins a steady decline as early as the mid-thirties, male fecundity rises more abruptly to a peak value during adolescence and then appears to be sustained as long as physiologically possible. The logic behind this difference in age patterns is tied to the much higher energetic costs of reproduction for females than for males and the associated trade-offs. Whereas a young female may pay dearly in terms of her own growth, survival and, ultimately, her lifetime reproductive success if she begins to reproduce too rapidly too soon, the male pays little cost for maintaining his physiological potential to mate when young even if opportunities are rare. Similarly, at older ages females face trade-offs between the benefits of additional offspring and the costs of additional reproduction in terms of their ability to survive and invest in offspring already born. As noted above, with increasing age the optimum balance may well shift for females in favor of lower fecundity. In males, however, the physiological costs of late age reproduction are unlikely to mount at anything near the female rate, while the benefits to lifetime reproductive success of fathering offspring late in life, especially if the mother is relatively young, remain nearly constant.

A third consequence also follows from the low direct energetic cost of reproduction for males. Male fecundity shows nothing like the exquisite sensitivity to energetic conditions that female fecundity shows. Studies of male marathon runners show no significant differences in sperm count, for example, between them and nonathletic men. Nor are increases in athletic training regimes associated with reductions in sperm production indices within individual men. Severe undernutrition, severe enough to threaten survival, can result in interrupted sperm production. At such extremes even marginal energetic savings may contribute to survival. But under the range of conditions humans normally face, including a broad range of energetic conditions, the cost of maintaining sperm production is relatively small. Even if limited energy availability reduces the probability that any offspring conceived would survive and flourish, the cost of making the attempt does not fall on the male. A female can actually lower her lifetime reproductive success by trying to reproduce under such circumstances, since substantial time and energy may be lost in an unsuccessful attempt. A male does not have to pay the same costs of failure. One might argue that in a pair-bonded species a male largely shares the reproductive success of his mate. Therefore he would also share the costs of impregnating her when the probability of a successful outcome are low. But if his fecundity were to be suppressed, he would also lose all possibility of impregnating other females as well. Even if such opportunities make only marginal contributions to a male's lifetime reproductive success, there would still be an advantage for a male to sustain his fecundity under conditions that lower his mate's fecundity. He can rely on his mate's physiology to lower her fecundity under stressful energetic conditions while maintaining his own fecundity in order to be able to take advantage of mating opportunities with other females.

To this point we have sidestepped one important issue: the difficulty of measuring male fecundity. In the discussion above, sperm production is the implied measure. Certainly anything that interrupts sperm production brings male fecundity to zero, but the significance of quantitative variation in sperm count is less clear. In clinical terms, sperm counts above 15,000 or 20,000 per milliliter of semen are considered "normal." There is little evidence that variation in sperm count above such a threshold increases a man's chance of fathering children, even though sperm counts can range to concentrations of 250,000,000 per milliliter or more. Perhaps the significance of such variation is not manifest in a monogamously mated couple trying to conceive. Perhaps, as with the fecundity of young adolescent males, the value of high sperm counts is more obvious when mating is more opportunistic. If, for example, it could be demonstrated for humans, as it has been for many domestic mammals, that a higher sperm count is associated with a higher rate of fertile matings, then the question of variation in male fecundity and its ecological correlates would need to be reexamined. Theoretically one might expect that at some point, however low the cost of sperm production, that cost could outweigh the diminishing returns derived from increasing sperm counts. An optimum level of sperm production might be affected by energetic conditions, but it might also be a very "soft" optimum, where large differences in sperm production make only small differences in fecundity. Although we cannot be sure, this description appears to fit what we know about normal variation in human sperm counts, with variation at the high end having little obvious relationship to male fecundity.

DRUG OF CHOICE

If the relationship between gamete production and fecundity is tenuous in men, the relationship between testicular hormone production and fecundity is even more so. Testosterone is necessary to maintain sperm production, but quantitative variation in testosterone levels within the very broad range of values recognized as normal does not appear to be related to quantitative variation in sperm production, either between men or within individuals. Yet testosterone levels do show predictable patterns of variation that seem to require some functional explanation. Free testosterone levels decline steadily with age, for example, and are consistently lower in men from traditional societies and rural communities in the developing world than among populations from developed countries. Although not as acutely responsive to moderate variation in energy balance and flux as female ovarian steroid levels, testosterone levels do seem to respond to long-lasting energetic conditions. Yet if testosterone levels themselves have no clear relationship to male fecundity, where do we look for a functional explanation of these patterns of variation?

We noted above that the basic constraints on male and female reproductive success are quite different. Because females carry the energetic burden of reproduction, female reproductive success is primarily limited by the availability of energy and the necessary time consumed by the process. Male reproductive success, by contrast, is primarily limited not by male fecundity but by mating opportunities. Energy can still be important for male reproductive success, but not the same way it is for females. If female reproductive physiology in mammals can be crudely characterized as a system for turning energy into offspring, male reproductive physiology can be crudely characterized as a system for turning energy into mating opportunities. It is in the management of this system that the functional significance of testosterone variation seems to lie.

There are two basic channels by which testosterone helps to manage the rate at which energy is converted into mating opportunities: through its effects on behavior and through its effects on metabolism. Steroids as a class of hormones are particularly suited to influence behavior owing to their ability to pass through cell membranes. This capability allows them to go where larger molecules like proteins cannot: across the blood-brain barrier, for instance. This barrier is composed of basement membrane that separates the brain and cerebrospinal fluid from the circulatory system. It allows the chemical milieu of the brain to be carefully controlled, but it also seals off the brain from many of the chemical signals that originate elsewhere in the body. The peripheral signal molecules that can most reliably reach the central nervous system are the steroids. Not only do they reach the brain, but it is clear the brain is listening. Many parts of the brain, especially in the limbic system and hypothalamus—areas of the brain concerned with the regulation of emotional and appetitive behavior and with the regulation of the anterior pituitary gland and all its downstream target tissues—are densely endowed with steroid receptors. Radioactively labeled steroids, injected into an animal's peripheral circulation, can soon be found bound to these central nervous system receptors.

Among the most obvious effects of gonadal steroids on the brain are their influences on reproductive behavior. Injecting an ovariectomized female rat with estradiol will elicit stereotypical behavior patterns that invite copulation from a male. Injecting a castrated male songbird with testosterone will elicit the singing associated with mate attraction and territorial defense. Numerous other examples can be cited to illustrate the same basic point. The behavioral effects of gonadal steroids help to coordinate reproductive behavior with reproductive condition. Animals do not typically display adult reproductive behavior patterns until reproductive maturation raises circulating steroid levels. In seasonally breeding species it is the renewal of gonadal activity (often linked to photoperiod or some other signal of the approach of appropriate breeding conditions) that stimulates the renewal of reproductive behavior.

We humans are, of course, complicated creatures, and the number of influences on our behavior are legion. Yet it is hard for even the most casual or biased observer to doubt the influence of gonadal steroid production on human reproductive behavior. Something transforms not only the bodies but the predilections of adolescents, and something else cools our sexual ardor with advancing age. Colloquially, parents refer to the influence of "raging hormones" on their teenagers, and pharmaceutical companies tout the boost to lagging sex lives of hormone replacement therapies for both men and women. But more direct evidence also exists to substantiate the claim that steroids affect several important aspects of human sexual behavior, and that these links are particularly important for males.

The traditional experiment used to demonstrate the influence of testosterone on male libido in experimental animals involves castration followed by exogenous testosterone replacement. Castrated male rats, for example, are much less likely than intact males to attempt copulation with a female in estrus. When given increasing doses of exogenous testosterone, castrated rats show an increasing probability of mating until at quasi-normal testosterone levels the castrated animals are just as likely as the intact ones to try to mate. This is not an experiment that is likely to pass the scrutiny of a Human Subjects Review Committee, or to draw many human volunteers if it did. Involuntary castration has been used, of course, to reduce the libido of slaves, prisoners, and the mentally retarded in different cultures at different times, and the use of chemical castration in the sentencing of repeated sexual offenders has been debated recently in our own society. But experiments very close to the rat experiments have, in fact, recently been conducted in the course of attempts to develop effective chemical contraception for men.

One of the prime obstacles to the development of a male "contraceptive pill" has been the link between testosterone and libido. The easiest way to interrupt sperm production pharmacologically is to eliminate or drastically reduce testicular testosterone production. This has a number of undesirable side effects, however, including the loss of libido, which would of course undermine the entire marketing value of the product. The idea, after all, is to enjoy sex without risk of conception, not to induce celibacy. Recently research has focused on ways to shut down sperm production and then exogenously replace testosterone to support normal libido. The first part of the goal—shutting off sperm production—is achieved by administering a long-acting GnRH analog. As discussed previously, such analogs mask the natural pulsatility of GnRH that is necessary for normal release of pituitary gonadotropins.

Without LH to support Leydig cell function and FSH to support Sertoli cell function, both testosterone production and sperm production come to a halt. It takes some weeks for all residual sperm to be eliminated from the testes and epidydimus, after which continued administration of the GnRH analog keeps gonadal function suppressed. Then exogenous testosterone is introduced, either through injection or transdermally from a patch worn on the skin. With appropriate controls, it is possible to study the relationship between the level of testosterone replacement and various measures of male libido, such as frequency of erections, sexual fantasies, masturbation, or intercourse. All of these show progressive responses to increasing testosterone doses until at normal testosterone levels the experimental subjects are no longer distinguishable from controls. There is no evidence that supranormal levels of testosterone produce exaggerated libido, however. Many factors may play a role in establishing an individual's sex drive as well as the translation of that drive into behavior. What the testosterone replacement studies do suggest is that testosterone is necessary to support normal male libido.

Interestingly, there is also compelling evidence that androgens are involved in supporting female libido as well. In women the major source of androgens is the adrenal gland, the gland that is also responsible for producing cortisol. Women who suffer from adrenal insufficiency are often treated with exogenous cortisol, but not exogenous androgens. When androgens are added to the treatment regime, the reported frequencies of sexual fantasy, activity, and pleasure all rise compared to the frequencies reported by controls treated with a placebo. Hormone replacement therapy for postmenopausal women usually involves exogenous estrogens and progesterone. But when androgens are added to that regime similar increases in reported frequencies of sexual fantasy, activity, and satisfaction are recorded.

As illuminating as these cases of androgen replacement are in demonstrating the role of androgens in supporting the libido of both sexes, they nevertheless represent abnormal situations. It is much more difficult to determine whether normal variation in male testosterone levels, either between or within individuals, is associated with variation in libido. Low testosterone levels associated with dramatic energetic stress, such as starvation or the combination of dietary restriction and intense exercise experienced by varsity wrestlers during training, are often associated with reports of low libido. But there are many other physiological changes confounded in these situations which make it difficult to confidently attribute the changes in libido to the effects of testosterone alone. One famous anecdotal account relating to this issue was published anonymously in the prestigious journal *Nature* in 1970. A single male researcher who was engaged in research on the Isle of Rhum off the Scottish coast for extended periods made a habit of collecting all his beard shavings from his electric razor each day and weighing them. This measure of daily beard growth provided him with a crude bioassay of his own testosterone levels. The report in *Nature* presented these data plotted day by day. Also indicated on the graph were the periods during which the researcher left the island to enjoy the companionship of his female partner on the mainland. The sinusoidal pattern of the beard growth data lends itself to interpretation as a graph of libido, lowest immediately after the researcher returned from the mainland and rising steadily in anticipation of the next visit.

Libido is not the only behavioral attribute that is credited to the action of testosterone, however. Aggressiveness seems to be associated with testosterone as well. Again the evidence is most convincing in animals. Castration has been used to reduce the aggressiveness of domestic animals for millennia. Testosterone replacement studies like those cited above show that restoring the testosterone levels of castrated male rats increases the probability that they will fight with other males in proportion to the dose of hormone received. Demonstrating similar effects in humans, however, is more difficult. In the testosterone replacement studies conducted in conjunction with the effort to develop a contraceptive for males, subjects did not report any noticeable change in their subjective sense of aggressiveness or anger. Geriatric patients who receive testosterone supplements to aid in the maintenance of muscle tissue report increases in positive feelings and elevations of mood, not any increase in irritability or tendency toward aggression.

It is possible, though, that we are searching under the wrong street lamp. It may be that testosterone contributes to a different set of behavioral attributes, one that *can* result in aggression under certain circumstances, perhaps more reliably in some species than in others, but that does not *necessarily* produce aggression in the absence of suitable provocation. It may simply be the dramatic nature of animal aggression and its unmistakable reduction following castration that attracted the attention of researchers and the label of a testosterone-mediated behavior. In fact, in humans available data do suggest a relationship between testosterone and something more like self-confidence or social assertiveness. For instance, in a study of male testosterone levels and occupations, James Dabbs of the University of Georgia and his colleagues found that while professional football players have higher testosterone levels than ministers, professional actors' levels are higher still! Cause and effect are difficult to determine in a study of this kind, but high testosterone levels have repeatedly been associated with occupations and lifestyles that involve risk-taking and assertiveness in social situations. It is easy to imagine that self-confidence and social assertiveness could lead to higher rates of aggression if the provocation is there. A bolder, more confident individual might be less likely to back down under a threat from another individual, and might even be more likely to engage in behavior that another could perceive as threatening in turn. In colloquial terms this kind of behavior is often associated with maleness and even male genitals. In American slang, a man with this set of characteristics might be described as "having balls," or an action that requires the expression of these traits as one that "takes balls." Similar expressions occur in many languages. But not all bold and assertive men are aggressive, nor are all aggressive men naturally bold and assertive. Human aggression rises from the confluence of many physiological, psychological, and sociological springs.

Self-confidence and assertiveness, however, are characteristics that can easily be associated with what in primate behavior studies goes by the name of social dominance. Dominance relationships among primates that live in social groups can usually be discerned from patterns of dyadic behavior that may or may not involve aggression. Dominance may be established between two males by the outcome of one or more aggressive encounters, and it may later be contested or even reversed by new aggression. But the majority of dyadic encounters between males reveal the relationship by the ability of the dominant male to assert himself and cause the subordinate male to retreat. The dominant male may assert himself at a food source, or claim proximity to a female, or simply approach a subordinate male and cause him to yield his sitting place. When the testosterone levels of monkeys are studied, the more dominant animals tend to have higher testosterone levels, although the correlation may not be perfect. Changes in dominance rank, however, are regularly accompanied by changes in testosterone levels in the same direction: the animal who goes up in rank goes up in testosterone while the animal who sinks in rank sinks in androgen level as well. In experimental situations where previously unacquainted male macaques are caged together, they quickly work out a dominant-subordinate relationship, with the subordinate animal undergoing a significant decrease in testosterone level and the dominant animal a significant increase. Similar patterns have also been observed among wild chimpanzees, our closest primate relatives, when dominance relationships change within a group.

Changes in testosterone levels of a comparable kind can be observed in human males following dyadic competitions in which there is a clear winner and loser. The first demonstration of this phenomenon was by a Harvard undergraduate, Michael Elias, using as subjects the members of the varsity wrestling team. He persuaded several members of the team to allow him to draw samples of their blood immediately before their matches, and again immediately after and an hour after the matches. Testosterone levels tended to be higher in all the wrestlers immediately after their matches compared to their pre-match baselines. In part this can be attributed to the physical exertion of wrestling, which

tends to shunt blood away from the liver and to temporarily reduce the rate at which hormones are removed from the circulation. However, the wrestlers who won their matches showed significantly greater increases than those who lost, and an hour after the match the losers' levels were significantly below their pre-match baselines while the winners' were still above theirs. Elias suggested that these different testosterone patterns were responses to the psychological impact of winning or losing.

This result has been confirmed and extended in numerous subsequent studies, many of which have sought to rule out possible alternative interpretations. For instance, it was pointed out that varsity wrestlers are often well acquainted with their opponents, either by prior experience or by reputation. Perhaps the losers expected to lose and hence didn't try as hard as they might have. Perhaps differences in levels of physical exertion often distinguish winners from losers and are also correlated with testosterone changes. The same qualitative results are obtained, however, even when wrestlers are matched against opponents they have never encountered or heard of. Nor is the phenomenon limited to contests involving physical exertion. Tournament chess matches also produce testosterone declines in the losers and rises in the winners. Even spectators, if sufficiently bound up in the vicarious experience of their team's fate, have been found to display similar changes in testosterone levels. Comparable changes have not been demonstrated in female competitors despite several investigations, a fact which simultaneously implicates testicular production as the source of the differences in males and undermines the possibility that they derive from changes in clearance rates of circulating hormones due to physical exertion. In one study, which involved collecting saliva samples from wrestlers every fifteen minutes after a match, the pulses of testosterone that are a recognizable reflection of testicular production ceased for several hours in the losers while continuing in the winners. This finding suggests that the divergence in testosterone levels between the contestants may be caused as much by the agony of defeat as by the thrill of victory. In this study the testosterone levels of winners and losers were once more indistinguishable three hours after the match. Recreational competition may have the advantage of transient effects. In a social situation such as a primate group where the same competitors may potentially renew their contest at any time, the effects may be longer lasting.

In the case of the outcome of human contests or primate dominance interactions, testosterone levels appear to be consequences of behavior, not antecedents. It is possible, of course, that testosterone can be both and thereby help provide a loop by which past experience affects future behavior. In a primate group, for example, it may be adaptive for an animal that suffers repeated defeats in agonistic encounters to adopt a less bold attitude, to assert himself less and give way to dominant animals more readily. If conditions change, however, and the same animal experiences a higher frequency of victories in unavoidable altercations, it may be to his advantage—particularly in terms of access to food and mates—to be bolder and more assertive. Shifts in testosterone levels may help to adjust these underlying behavioral tendencies without predetermining an animal's response to any particular situation. Similarly in humans, testosterone changes may reinforce the shifts in self-confidence that can follow significant defeats and victories. There is already speculation that this phenomenon may contribute to slumps and streaks in athletic performance. It is tantalizing to imagine that testosterone responses may play a similar role in the social world of human males as well, where competition may be more subtle but just as intense as in the wrestling ring.

BE ALL THAT YOU CAN BE

In addition to the behavioral effects of testosterone that may be of importance to male reproduction and the direct physiological effect of supporting sperm production, testosterone also has important physiological effects in promoting the production of muscle mass. This effect of testosterone has gained a certain notoriety through the use and misuse of anabolic steroids by athletes. Testosterone is the original anabolic steroid (anabolism being the term for "building up" molecules, as opposed to catabolism, the term for "breaking down" molecules). Testosterone has been used directly as an anabolic hormone to promote muscle development in both male and female athletes and body builders. It has also been used to help geriatric patients maintain muscle mass and the strength and balance that come with it. Other anabolic steroids have been developed synthetically in an attempt to maximize muscle-promoting effects while minimizing other masculinizing effects, as well as to avoid detection in sports where steroid use is illegal. Controversies surrounding the use of anabolic steroids by athletes, especially in sports such as track and professional football where they are prohibited, have surfaced repeatedly. The East German women's Olympic team was criticized by other nations for using them, Ben Johnson was stripped of his Olympic gold medal for steroid use, and Mark McGwire drew censure when he acknowledged that he had regularly used androstenedione, a weak androgen of debatable effect, during his record-breaking home run season. Few knowledgeable observers doubt that National Football League linemen regularly use anabolic steroids to achieve bulk, strength, and speed. Nor is there much doubt of their regular use in sports where they are not regulated, such as professional wrestling and body building.

Testosterone supports muscle building by promoting the uptake of glucose and amino acids by muscle cells and their use in producing structural protein. This action can be contrasted with the action of estradiol in promoting the uptake of glucose by fat cells. Prior to puberty boys and girls have quite similar amounts of muscle and fat on their bodies. But at puberty rising levels of the dominant gonadal steroids, testosterone in boys, estradiol in girls, send their metabolisms in different directions. Boys continue to accumulate fat in absolute terms, if ecological circumstances allow, fat that they need to ensure their own survival in case of insufficient food. But excess energy is primarily directed toward muscle growth. Girls continue to build muscle as they grow, but excess energy is primarily directed toward storage as fat. The resulting divergence of body composition is a consequence of both of these processes. In absolute terms, the average adult male carries a similar amount of fat on his body as the average adult female, but his muscle mass may be twice as great. His frame is also larger on average. But even when we correct for differences in overall size, the large difference in body composition between males and females is more a function of differences in muscle mass than of differences in fat mass.

The importance of testosterone in mediating changes in muscle mass can be vividly demonstrated by data derived from studies in which researchers used GnRH analogs to halt endogenous testicular hormone production and then later replaced the testosterone exogenously. Men who are deprived of their own endogenous testosterone in this way show decreases in protein production and protein turnover and loss of muscle mass and strength. At the same time, they show increases in fat deposition. Replacing the testosterone reverses these effects and leads to increases in muscle mass and strength. Similarly, exogenous supplementation of testosterone levels in older men leads to increases in muscle mass and strength and decreases in fat mass.

Of course, testosterone alone doesn't produce muscle. Two other elements are important. Energy must be available to convert into muscle, and the muscle must be used. A man who is starving will begin to catabolize his own skeletal muscle as a source of amino acids and energy to stay alive. In order to add to one's muscle mass there must be energy available to devote to the task after other metabolic demands are met. Not only does muscle require energy to build, unlike fat it requires energy to maintain as well. Skeletal muscle can account for 20 percent or more of a man's basal metabolic budget, which is not a trivial fraction. Ounce for ounce, muscle is not as expensive to maintain as some other metabolically active tissues, such as brain, liver, or kidney, but it is more expensive to maintain than inert tissues such as bone or fat and some metabolically active tissues such as skin. More important, it is the metabolically active tissue that is most expendable and adjustable. It is not very feasible—nor would it be adaptive—to reduce one's brain size or kidney size or skin coverage to cut metabolic costs if energy is in short supply. It is, however, possible to reduce muscle mass under adverse conditions and to build it up again under favorable ones. The mechanisms that link muscle anabolism to energy availability are simple. Muscle cells take up glucose in response to insulin secretion, which in turn is determined by energy availability. Testosterone acts to enhance this process, in part by contributing to the expression of insulin receptors by the muscle cells.

Flexibility in muscle mass and its consequences for the overall metabolic budget relate to the other requirement for building muscle: the muscle must be used. If you are planning to take anabolic steroids regularly and then sit back and wait for your biceps to bulge, you are in for a disappointment. You still have to go to the gym and do biceps curls if you want results. Use of the muscle stimulates the production of the enzyme systems necessary to build more contractile fibers and other functional elements of the muscle cell. Testosterone enhances this response but it can't create the response where it doesn't exist. Disuse of a muscle will send things in the other direction, causing atrophy of muscle tissue and loss of strength. Even the most muscular man will find his leg muscles wither appreciably if he has his leg in a cast for a month or more.

The anabolic response of muscles to habitual use is muscle specific as well. Exercising the biceps will not produce massive quadriceps. Wheelchair marathoners simultaneously illustrate the effects of use and disuse. Their upper bodies are typically heavily muscled while their leg muscles have often atrophied completely owing to disuse resulting from paralysis. The body seems to adjust muscle mass rather naturally to meet the demands of habitual use. "Use it or lose it" is a principle that applies nicely to muscle in the interests of metabolic economy. Alas, if only the same principle applied to fat!

Testosterone, then, does not cause muscle to grow by itself. It enhances the body's ability to accomplish this task when conditions are right, when energy is available, and when habitual use indicates that additional muscle mass would be functional. When men are matched for diet and exercise but given different doses of exogenous testosterone, higher testosterone results in faster muscle growth and greater eventual muscle size. Differences in absolute testosterone levels between individual men do not always predict muscularity, however, since testosterone alone does not determine muscle growth. Men may differ in their exercise habits, their diets, or even their normal testosterone receptor density in muscle tissue. As in the winners and losers of wrestling matches discussed above, the changes in level within individuals are of greater significance than the absolute levels. Large differences in absolute levels of testosterone, such as that between males and females, clearly make a difference, however.

Yet while the average difference in muscle mass between males and females is not particularly controversial, the functional reasons for the difference are more likely to be. One common assumption is that males are biologically endowed with a greater propensity to accumulate muscle mass because the tasks they perform under the traditional sexual division of labor require more strength. They must perform the heavy work, it is argued, while women are responsible for lighter work such as child care and food preparation. If such a division of labor has been typical of human societies for the majority of our evolutionary history, the argument continues, men will have been selected for greater muscle mass. This characterization of the traditional division of labor often seems bizarre to those who have spent time with hunter-gatherers or subsistence agriculturalists. For example, Lese men in the Ituri Forest are responsible for clearing new gardens for planting, an activity that requires strenuous labor for several weeks. During the rest of the year, however, they can often be found resting in their hammocks while their wives carry firewood and water weighing nearly as much as themselves for long distances. Moreover, to the extent that labor tasks are divided between the sexes on the basis of the strength required, it is as likely for such division to be the result of differences in muscle mass as it is for it to be the cause.

But perhaps the most telling observation is that differences in muscle mass are typical of many species where sexual division of labor is not an issue. In many other primates, for example, including common chimpanzees *(Pan troglodytes)*, gorillas, baboons, and macaques, males have significantly more muscle mass than females. Features of this type, characteristic of one sex but not the other, are often the result of sexual selection; these features contribute to the ability of individuals to compete with members of its own sex for opportunities to mate with the opposite sex. The feature may be used to attract mates, like the elaborate tail display of the peacock, or to fight with rivals, like the antlers of the deer. Very often both functions are served, since the other sex may find features attractive that provide the bearer with a competitive advantage in direct struggle with members of its own sex. In cases of sexual selection among mammals it is most often males that are competing for access to females for the reasons discussed above, and primates are no exception. The muscle mass of males and the formidable canine teeth they often bear are used primarily in confrontations with conspecific males, confrontations that can at times result in death. The keener the rivalry for mating the more exaggerated the development of these primary weapons. Where the rivalry is relaxed by ecological circumstances, as among the pygmy chimpanzees, or bonobos *(Pan paniscus)*, sexual dimorphism in muscle mass is also dramatically reduced.

It is likely that the sexual dimorphism in muscle mass displayed by humans is one more example of sexual selection for competitive ability vis-à-vis other males. Perhaps males do not often come to blows over their prospective mates in the modern world (though this may happen more than we realize, as homicide statistics suggest), and perhaps women do not necessarily find muscular male physiques attractive (though the film and advertising industries are betting otherwise). It is easy to imagine, however, that physical strength may well have been important for male social dominance and mating opportunity in our evolutionary past, as it apparently is for our primate relatives in the present.

MAKING AN EFFORT

In previous chapters I have argued that variation in ovarian function helps to modulate female reproductive effort in humans, using the term "reproductive effort" in its technical sense. In females, reproductive effort primarily takes the form of physiological investment of time and energy in the production and nursing of offspring. A similar argument can be made that variation in testicular function helps to regulate male reproductive effort. For males, reproductive effort primarily takes the form of the investment of time and energy in gaining mating opportunities. Some of this investment is behavioral, some physiological. Testosterone contributes to both pathways by helping to modulate libido, or the motivation to pursue mating opportunities; by promoting self-confidence and the social assertiveness necessary to compete effectively for mating opportunities with other males; and by helping to build muscle mass, which enhances male competitive ability. Increases in testosterone within an individual appear to enhance all these elements of male reproductive effort, while decreases lower them. Sperm production remains largely unaffected since it is not the capacity for fertilization that is costly or limiting in terms of male reproductive success, but rather gaining the opportunity to mate.

In the light of this understanding of testosterone and its effects, we can examine patterns of variation in male testicular function and their relationship to ecological and constitutional factors. We should note at the outset that that there is tremendous variation between individuals in testosterone levels, as much as a tenfold range, all of which is considered "normal." As with the variation in sperm counts noted above, the significance of these baseline individual differences in testosterone is not clear. Clinical research tends to treat variation in this broad "normal" range as inconsequential, concerning itself only with "abnormal" testosterone levels. From this perspective there is little notable variation in testicular function among populations of healthy males since there is little variation in the frequency of abnormal testosterone readings. Among twenty-nine Lese and Efe males from the Ituri Forest, for example, only one had a testosterone level that fell below the normal range. As a group, however, the Ituri males had average testosterone levels that were significantly lower than males of similar age in Boston had. These differences were particularly great in the morning, when testosterone levels tend to be high. Studies of a number of traditional societies, including Tamang and Kami men in Nepal, !Kung men in Botswana, Turkana men in Kenya, Aymara men in Bolivia, and Aché men in Paraguay, also document testosterone levels that are significantly lower on average than observed among populations in developed countries, although the majority of individuals in all these populations fall within the "normal" range. As with ovarian function in women, then, testicular function in men, indexed by testosterone levels, appears susceptible of quantitative variation that is not pathological. Testicular function can be higher or lower, not just on or off.

The causes of this quantitative variation in average testosterone levels between populations are more difficult to determine. Variation in energy expenditure does not seem to be associated with reduced testosterone levels. Indeed, given the role of testosterone in supporting increases in muscle mass it would be odd if increases in habitual muscle use resulted in lower testosterone. Neither does moderate variation in energy balance appear to lower testosterone levels. Among the Lese, for instance, the seasonal weight loss that produces significant changes in female ovarian function doesn't appear to have a similar effect on male testosterone levels. Nor do Tamang males in Nepal show any correlation between testosterone levels and energy balance comparable to the variation in ovarian function documented among Tamang women. Chronic energy shortage may be a different matter, however. All the populations mentioned above that have low average testosterone levels compared to Western populations also show evidence of chronic energy shortage both in terms of low levels of stored fat and in terms of their short stature as gauged by international growth standards. It would make sense, physiologically, for testosterone to correlate with long-term but not short-term energetic conditions. Short-term variation in energy availability may result in mobilization or storage of fat in males as it does in females. It would not be very adaptive to meet with such short-term variation by alternately breaking down and building up muscle. Not only is it a much more expensive process, metabolically, than the drawing down and building up of fat reserves, but it also has appreciable effects on physical performance. Longer-term energy shortage, on the other hand, may require a lowering of metabolic expenditures. Reducing muscle mass may be a step toward meeting such challenges. Long-term positive energy balance may similarly provide an opportunity to increase muscle mass and competitive ability. By this logic testosterone should not show the sensitivity to acute energetic conditions that female ovarian function displays, but should track longer-term energetic conditions instead.

Whether we should also expect permanent developmental effects of energetic conditions on testicular function similar to those postulated for ovarian function in females is less clear. For females, we noted that the metabolic costs of reproduction remain fairly inflexible. A small woman must invest nearly as much in the gestation and nursing of her offspring as a large woman, although these costs may represent a greater fraction of her total metabolic budget. Chronic energy shortage that results in small adult stature may require women to space births more widely to maintain long-term energy balance and optimize their reproductive effort. Therefore we expect chronic energetic conditions that result in slow growth and late maturation also to result in lower average female fecundity and ovarian function. In males the situation may be somewhat different. Smaller adult stature may allow a male to maintain a higher percentage of muscle under conditions of energy shortage. A smaller man may be able to support more muscle than can a taller man under the same energy budget, and the same absolute amount of muscle will be a greater proportion of total mass on a smaller man. Reducing adult stature under conditions of chronic energy shortage may be part of an overall somatic strategy that allows a man to optimize his reproductive effort as well as his survival.

Age patterns of testicular function appear to be quite different from age patterns of ovarian function. For one thing, there is no equivalent to menopause in men. We noted in the previous chapter that female menopause is a consequence of a finite supply of oocytes. Male spermatocytes are continuously replenished by mitosis from the germinal epithelium of the semeniferous tubules, so there is no inevitable moment at which the supply runs out. That is not to say that testicular function shows no effects of age. Virtually every index of testicular function declines with age in older men: sperm counts, sperm motility, testosterone levels, and, most notoriously, the ability to achieve and maintain an erection.

Many of the indices of testicular function that decline with age appear to be correlated with general health and physiological robustness. In this respect declining testicular function looks very much like another manifestation of general senescence. Men who remain healthy and physically vigorous into old age are more likely to maintain testicular function at higher levels than men whose general health declines more rapidly. Given this fact, it has been difficult to determine whether there is any underlying age-related decline in testicular function that is independent of general senescence as the decline of ovarian function prior to menopause appears to be in women. Testosterone is a promising index of testicular function to assess, since the relevant data are easier to collect from large random samples than data on sperm count or potency, for example. There are problems to overcome with testosterone as well, however. For one thing, total testosterone, the measurement usually made in blood, may be misleading since it combines a measure of free, biologically active hormone with a measure of protein-bound, inactive hormone.

Salivary testosterone provides a very good index of free testosterone since only free testosterone passes through cell membranes to enter the saliva, and since it does so in near perfect equilibrium with free testosterone levels in the blood. Levels of salivary testosterone in samples of men from the Boston area screened only to exclude illness or steroid medication decline quite steadily, nearly linearly in fact, from early adulthood on. The range of individual variability is, as noted previously, quite high, but the downward trend is highly significant and quite steep. Levels of salivary testosterone at the age of sixty are on average less than half of the levels at age thirty. In part this decline may reflect changes in the ratio of bound to free testosterone. But the frequency of testosterone pulses declines with age as well, suggesting that testosterone production is also reduced.

Salivary testosterone generally appears to decline quite linearly with age in other populations as well, although the slope of the decline can vary a great deal. In some populations, however, the slope is so flat that the decline is no longer significant. When this occurs it appears to be associated with low testosterone levels in young adulthood, not high levels in old age. The Aché in Paraguay and the Tamang in Nepal exhibit this pattern, for example. Comparing the age patterns of testosterone for Aché, Tamang, Lese, and Boston men indicates that the higher the early adult testosterone level, the steeper the decline with age. This suggests that the decline in testosterone is more a function of variation in youth than variation in old age. Testosterone levels in older men are much the same regardless of the population from which they are drawn. Rather it seems that early maturing populations may reach levels of testosterone early in adulthood that are difficult to sustain. It remains unclear whether the linear decline from such high early adult levels is in fact inevitable, or is merely a correlate of declining physical activity with age in Western populations.

Thinking about testosterone in terms of the regulation of male reproductive effort may shed some light on the question of male growth and reproductive maturation left unresolved a few chapters earlier. Presumably there is no fixed critical size for males to attain to be reproductively successful. Other things being equal, bigger is better *if* bigger translates into superior competitive ability. Environmental constraints in terms of chronic energy availability, however, mean that bigger is not always better. For given environmental circumstances there is an optimal size allowing enough muscle to be maintained to enhance competitive ability without too great a sacrifice of survival probability. It is in a male's favor to grow as quickly as can be managed to this size and then to stop, putting available metabolic energy into muscle mass and the behavioral components of reproductive effort. Under more stringent environmental conditions growth will be slower and final adult size smaller. Under more favorable conditions growth will be faster and final adult size greater. Because males are trying to be "as big as possible" rather than "big enough" they grow for a longer time than females, two to three years longer on average, before shifting metabolic resources from growth to reproduction. Male stature is also more environmentally variable than female stature. Under the conditions of virtually unlimited energy availability that privileged populations enjoy in the modern world, growth appears to reach the limits of its flexibility. If, however, men were still to compete physically for mating opportunities, the stage would be set for the evolution of even greater male size and strength.

There is much more that we need to know about the ecology of testicular function. But what we do know of that ecology seems to make sense as a mechanism to modulate male reproductive effort. The key is to realize that male reproductive success is tied to competition for mating opportunities. Testosterone can be understood as a major regulator of male reproductive effort since it helps to modulate the motivation to seek mating opportunities (libido), the motivation to compete socially with other males (self-confidence and social assertiveness), and the allocation of somatic resources to make such competitive engagement effective (muscle mass). Chronically favorable energetic circumstances enhance all of these through raising testosterone levels, whereas chronic energy shortage reduces them. Peak reproductive effort, as reflected in peak testosterone levels, occurs soon after the attainment of physical maturity and declines as physical vigor declines. Other things being equal, young men have higher testosterone (and hence are expected to have greater reproductive effort) that older men; men under chronically favorable energetic circumstances have higher testosterone than men under unfavorable circumstances; healthy, vigorous men have higher testosterone than do sick, inactive men. In general, however, all men maintain sperm production as long as possible through all but truly life-threatening emergencies, since the energetic requirements of sustaining fecundity are so low. Hope, it seems, dies hard in the gonads of the human male.

THE JOURNEY AND THE PROCESSION

THE MAIN ARGUMENT OF THIS BOOK has been relatively simple: viewing human reproductive physiology as a product of evolution, responsive to ecological conditions, helps us understand why it works the way it does. But it's also possible to combine this understanding with other information in an attempt to clarify the process and history of human evolution itself. Ever since Darwin, the mystery of human evolution has been particularly intriguing. How did an animal something like a chimpanzee give rise to a creature so smart and talented that it often has trouble thinking of itself as an animal at all? What selective pressures were involved? What were the steps along the way? The evidence usually brought to bear on these questions comes from a variety of sources: fossil evidence of anatomical changes, archaeological evidence of past behavior, genetic evidence of similarity to and difference from other primates, behavioral comparisons with other species. What happens if we add our understanding of the ecology of human reproduction to this mix?

THE JOURNEY

The outline to the story of human evolution is reasonably clear. Sometime around five to seven million years ago the hominids—the line of descent leading to modern humans—branched away from the line giving rise to the modern African chimpanzees. We infer this timing from genetic evidence primarily, but it fits the accumulating fossil evidence so far as well. The two chimpanzee species, the common *(Pan troglo-*

dytes) and pygmy (*P. paniscus*, or bonobo) chimpanzees, are very closely related to each other, and probably separated from each other after the hominid line split off. Despite surface appearances that long caused taxonomists to group chimpanzees and gorillas together, separated from humans, the vast weight of genetic evidence now indicates that humans and chimpanzees are more closely related to each other than either is to gorillas. Or put another way, humans and chimpanzees share a common ancestor that is more recent than the common ancestor shared by gorillas and chimpanzees. This fact together with the observation that chimpanzees and gorillas are very similar morphologically allows us to infer with reasonable certainty that the first hominids probably started off morphologically similar as well. The common ancestor of modern chimpanzees and humans probably lived in the rain forests of central Africa, lived on a diet of fruit supplemented by a variety of other plant materials and occasionally meat, and had a breeding pattern that was not seasonally restricted. Interbirth intervals were probably quite long; offspring were born singly and remained with the mother for a number of years, possibly until reproductive maturity between the age of six and ten.

From that beginning, hominids spread out into the African woodlands, scrub, and savanna. By three million years before the present there were several species of hominids, differing apparently in size and diet, resident along the eastern side of Africa from modern Ethiopia to South Africa. By two and a half million years ago some of them began to make stone tools. There is clear evidence of meat eating. Bones of small antelopes have been found with cut marks from stone tools where they had been jointed and butchered. Fossilized footprints across the drying mud flats of ancient lakes provide indisputable evidence of well-developed bipedal locomotion in some hominids. Tooth size varies appreciably among these early hominid species, suggesting differences in diet. Some with massive molars probably specialized in grinding quantities of coarse plant material. Others with smaller teeth probably depended more on fruit and meat. High-resolution electron microscopy reveals patterns of scratches and striations on well-preserved fossil teeth that match these dietary differences. The great, flashing canines of our primate past had disappeared by this time, however, reduced in size to more closely match the adjacent incisors and premolars. The jaw still jutted out in the front of the face. The brain case was still quite small, no bigger than a chimpanzee's in some species, half again as big in others. Some researchers give some of these creatures the genus name *Homo*, indicating the working hypothesis that they are closer to the line leading to modern humans than their contemporaries, who bear the genus name *Austalopithecus*.

A bit less than two million years ago a new species of the genus *Homo* appears, *Homo erectus*. There is little doubt that this creature is a direct ancestor of modern humans. It is larger that previous hominids, with a brain more than twice the size of a chimpanzee's, though still only half the size of our own. It uses fire to cook its food. Its range extends beyond Africa to Europe and southern and eastern Asia and into habitats much different from the African savanna and woodlands that were home to its ancestors. It makes more sophisticated stone tools, and possibly clothing and shelters as well. By the end of another million years it is the only species of hominid on the planet. The reasons for the disappearance of the others remain unclear.

Slightly less than half a million years ago our own species emerges. The first evidence of *Homo sapiens* again appears in Africa. This species replaces *Homo erectus* over its entire range within a few hundred thousand years. Debate continues over the nature of this replacement, whether it represents one or more waves of migration of *H. sapiens* out of Africa, or a gradual evolution in situ throughout the range of *H. erectus*, though majority opinion favors some version of the former scenario. Eventually this species expands into lands never reached by *H. erectus*, including Australia and the Americas. It comes to inhabit the broadest range of habitats of any animal ever to live on the earth. The brain of *H. sapiens* is half again that of *H. erectus* in size. With the larger brain comes evidence of greater intelligence in many realms, more sophisticated tools, art, clothing, shelters, cooperative hunting, ritual burials, and music. Anatomical changes in the vocal tract together with abundant evidence of complex behavior strongly suggest fully developed language ability.

Piecing together the story of human evolution is very much like working on a jigsaw puzzle with loosely fitting pieces, the majority of which are missing, without knowing what picture is on the cover of the box. The picture emerges slowly as new pieces are discovered and placed. Sometimes the new pieces lead to a rearrangement of old pieces and the emerging picture changes. Sometimes gaps are filled in that add new information in the light of which the whole picture is reinterpreted, if not rearranged. From our consideration of human reproductive ecology we can place a few more pieces on the table, not all new ones necessarily, but ones that are often overlooked. Somewhere along the way birth became difficult and social support at delivery became necessary. The pelvis became sexually dimorphic, reflecting the obstetric constraints on females. At some point both males and females began to live routinely past the age at which female menopause occurs. At some point birth intervals became shorter and reproductive maturation took longer. It became frequent for women to have more than one dependent offspring at a time. At some point the body substantially increased its potential for storing fat. Although we can't be sure yet where these pieces fit, we can try to arrange them with the others as best we can to see what they might add to our understanding of the picture.

Of central importance in the story of human evolution is the brain. This is clearly what sets us apart, the thing that adds to our physical capacity to hold things and make sounds the cognitive capacity to design bridges and discuss politics. To be sure, other primates are quite intelligent, chimpanzees especially so, and the more we learn about their cognitive abilities the more it seems that their intelligence is not so much different in kind as in degree. But of that difference in degree there can be no doubt. Making primitive stone choppers may not represent a tremendous advance over fashioning a stick to catch termites or a leaf sponge to collect water, as chimpanzees are known to do. Chimpanzees in the Tai forest are known to use rocks to open hard nuts. It is not too difficult to imagine that a change in habitat to open savanna, together perhaps with a change in diet that placed a premium on ways to split open scavenged bones to get at marrow, might lead to greater use of stone as a tool material by an early hominid with no greater intelligence than a chimpanzee. But the gulf from that starting point to the mental prowess of *Homo sapiens* is immense.

What was it that provided the driving force for the evolution of the human brain? One might imagine that increased intelligence is generally adaptive, that any animal with an advantage in brains over its fellows might use that intelligence to survive and reproduce with greater success. But such a general advantage to intelligence can't be the whole answer, since only humans have taken this particular evolutionary journey, or taken it so far. Nor is it clear that intelligence is one "general" thing rather than a collection of particular mental abilities pertaining to different sorts of challenges. Grappling with this problem has led to much speculation about the particular selective pressures for intelligence that may have been peculiar to human ancestors. Perhaps it was the complexity of their social organization that selected for particular mental traits, or the way they acquired their food, or their lack of physical defenses against predators. These sorts of arguments are difficult to elaborate convincingly, since the hypothesized selection pressures are rarely unique to humans. A common variation on this theme is to invoke a snowballing effect of intelligence selecting for its own advancement. A more intelligent creature creates a more complicated social world in which more intelligent creatures are more successful, and so on. But why this snowball rolled so much farther for hominids than for other apes is left unexplained.

A different approach, taken by some, focuses less on unique selective pressures that might drive brain evolution and more on the removal of constraints that might impede it. The primary constraint appears to be the metabolic cost of brain as a tissue. The high and nearly continuous level of metabolic activity of the brain makes it one of the most expensive tissues in the body to maintain, along with the heart and the digestive tract. The brain is like a large household appliance that is left on all the time, continuously drawing power. Leslie Aiello and Peter Wheeler note that because animals of a given size tend to have similar overall metabolic rates, having a big brain relative to body size requires a proportionate decrease in either the amount or the metabolic activity of some other tissue to compensate. The tissue that seems most often to be involved in this trade-off in primates is the digestive tract. The bigger a primate's brain is for its body, the smaller its gut. Conversely, bigger than average guts seem to be associated with smaller than average brains. Big guts are associated with diets that involve consuming large amounts of food with low caloric content, often food that is difficult to digest, such as plant material heavy in fiber and cellulose. Gorillas, for example, have large digestive tracts and subsist mainly on herbaceous vegetation. Maintaining such a large gut requires more of their metabolic energy than is true of the average primate. Having a relatively smaller brain helps them compensate for this expense. Chimpanzees, with a diet rich in fruit, have much smaller digestive tracts for their body size than gorillas, and relatively bigger brains.

Several researchers have drawn from these observations the obvious conclusion: that having a large brain requires having a high-quality diet, one based on food of high caloric density that is easy to digest. A dietary shift becomes the key element in human evolution by this way of thinking, opening the door for the evolution of an outsized brain and off-the-scale intellectual capacities. The favorite item to point to in the human diet that seems to set it apart from other primate diets is meat. Other primates eat meat, but not as a main part of their diet. Meat is relatively high in calories, and where there is meat, there often is fat, in the subcutaneous depots, internal organs, and bone marrow of animals. Both animal fat and meat are relatively easy to digest and would have become even easier to digest with the introduction of fire and cooking.

There are a couple of reasons, however, to think that meat eating is not the full answer to the riddle of human brain evolution. On the one hand, there are entire groups of mammals that specialize in meat eating, large carnivores like the big cats, wolves, dogs, and hyenas, and smaller carnivores like weasels and civets, none of which displays exceptionally large brains or human-like cognitive abilities. On the other hand, meat is a scarce item in the diets of millions of people around the world, particularly in developing countries. Many of these people may suffer from malnutrition is various ways, and may even suffer measurable cognitive deficits, but there is no question of their ability to support human-sized brains and to display human levels of intelligence. Agriculture provides an alternative to meat in cereal crops that are relatively rich in calories and, when cooked, relatively easy to digest. It may be that meatless human diets became viable only with the advent of this alternative source of staple foods. Nevertheless, a meat-based diet seems neither necessary for the maintenance of a large brain nor sufficient to drive its evolution.

One crucial constraint on brain evolution is overlooked by all these arguments: not only are large brains expensive to maintain, they are expensive to grow. The burden of this expense must primarily be born by the mother during pregnancy and lactation. The human brain grows dramatically in utero and continues to grow rapidly during the first two to three years after birth. Although a human has a gestation period only a month and a half longer than that of a chimpanzee, the newborn human has a brain that is twice as large. Yet the human brain is still only half its final size at birth, while the chimpanzee brain is nearly full-sized. As a result, human infants are still very underdeveloped neurologically at the time of birth and won't match a newborn chimpanzee's level of sensorimotor ability and coordination for a couple of years. We have seen evidence of the physiological priority placed on this brain growth during gestation, with the mother drawing down fat reserves laid up earlier in gestation in order to meet the ever increasing metabolic burden on infant brain growth. We have seen evidence of the selection pressure on the female pelvis to accommodate the passage of a larger fetal head. We have noted the importance of the shift to lactation as a means for the mother to continue to meet the nutritional needs of her infant from her fat stores, and the evidence for the metabolic burden that lactation represents. It may be that the most important constraint in the evolution of the human brain was not strictly nutritional but reproductive; not a question of how to maintain a large brain in an adult, but how to grow a baby with one.

The ability of human mothers to support the growth of their large-brained babies seems to be based on their ability to store and mobilize large amounts of fat. But where did this ability come from? As we have noted, both male and female humans have a capacity for fat storage that is greater than that of many other primates, including chimpanzees. Fat storage capability of this kind usually evolves in mammals when food availability undergoes periodic shortages that cannot be fully circumvented by switching diets. For example, herbivorous mammals living at high latitudes usually have considerable fat storage ability in order to survive through the winter season when plant foods are scarce. Their relatives in low latitudes do not store fat to anything like the same degree. At least one ape shows a similar disposition to store fat when times are good and then to live on those reserves when times are bad: the orangutan. Cheryl Knott has shown that orangutans in Borneo live in a forest that has dramatic swings in fruit availability due to the synchronous flowering and fruiting of many of the key tree species, a phenomenon known as masting. At some seasons fruit is available in great abundance and all other items in the orangutan diet diminish in importance. At other seasons, after the mast, fruit is exceptionally scarce and orangutans fall back on foods of only marginal nutritional content such as bark. At these times levels of ketones rise dramatically in their blood indicating a high rate of fat catabolism. Chimpanzees in Uganda undergo shifts in fruit availability, but they are not nearly so dramatic as those in Borneo and better fall-back foods are often available. Knott has also uncovered evidence of changes in orangutan ovarian function that parallel the shifts in fat storage and mobilization caused by the variation in fruit availability: orangutans have significantly higher levels of ovarian function during times of positive energy balance and low levels during times of negative energy balance.

It seems likely that hominids developed an enhanced fat storage ability as a response to a similar ecological pressure, a pattern of food availability that was characterized by alternating periods of abundance and dearth, selecting for the ability to store and later mobilize energy in large quantities. Perhaps this change accompanied the change in habitat from rain forest to wooded savanna and a shift in diet away from fruit toward a diet based on tubers and herbaceous plants with unpredictable bonuses in the form of scavenged carcasses. Taking full metabolic advantage of these sporadic dietary windfalls may have been the key to survival. Locating carcasses scattered broadly across the African plain probably required an ability to cross long stretches of open ground. This may have increased the selection pressure for efficient bipedal locomotion. Gaining access to carcasses once they are found might have involved hurling things at competing scavengers to drive them away. Getting meat and marrow from the carcasses might have involved manipulating stones. Both of these activities could have increased the selection pressure for manual dexterity and tool use. But fully utilizing the metabolic energy represented by a carcass would have involved the ability to store much of it as fat.

Not all hominid species on the African savanna necessarily took this route. Some may have developed different survival strategies. One alternative may have been to increase in body size and gut capacity and to develop a dentition suitable for processing large amounts of coarse plant material, becoming more of a hominid cow, if you will, or a savanna gorilla. This metabolic strategy might be successful, but would lead away from the potential for increased brain size toward a smaller-brained, larger-bodied animal. Some of the hominids that inhabited the savanna two to four million years ago appear to have taken this path. But at least one hominid species appears to have remained smaller, developed dentition more suitable for softer food, and begun to increase its brain size. This strategy may also have involved an enhanced ability to store and mobilize energy reserves in order to smooth out booms and busts in food availability.

The increase in brain size was modest at first, but may well have facilitated increasingly sophisticated tool use and perhaps an increasingly complex social life. But brain size was still close to the norm for a primate of equivalent size, and there was no evidence of dramatic pelvic dimorphism and no reason to suppose that birth was particularly difficult. Bipedality does seem to have selected for changes in pelvic shape that may have made it necessary for a full-term fetus to rotate its head somewhat during delivery, but passage of the fetal head was probably still relatively easy.

The increased ability to store metabolic energy may have set the stage for greater increases in brain size, however. In the terminology of modern evolutionary theory, this ability to survive periods of low food availability may have been a preadaptation (or exaptation) for the ability to support greater fetal brain growth. A shift in subsistence ecology that made the flow of dietary energy more predictable and steadier, though not necessarily constant or superabundant, may have allowed fat storage ability to be put to this new use. Perhaps a behavioral shift from scavenging toward active hunting was involved. Perhaps increasingly sophisticated tool use led to the utilization of food sources that were previously inaccessible. Perhaps scavenging itself became a more reliable way of life as tool use and coordinated social behavior increased. A more reliable flow of energy would have eased the pressure on energy storage for survival and freed up energy for other metabolic tasks, such as increases in body size and muscle mass in males. In females, the ability to accumulate and mobilize fat efficiently may have found new utility in supporting the growth of a larger-brained offspring. As noted earlier, a pregnant woman uses essentially the same metabolic processes to mobilize her reserves during late pregnancy that she would use during a famine or fast, leading to the characterization of pregnancy as a state of "accelerated starvation."

A shift of this type may have been involved in the emergence of *Homo erectus*, a hominid with larger overall body size but a markedly enlarged brain relative to its body. With the increase in brain size apparently came an increase in brain power, significant technological advances in tool manufacture and the use of fire, and an ability to invade and exploit new habitats and to expand beyond Africa. Now the snowball may really have been set rolling. With technological and cultural advances the flow of metabolic energy may have become even more dependable. Birth intervals may have begun to shorten, leading to population expansion concomitant with geographic expansion. Even greater fetal brain growth may have become possible until the limits of maternal energy transfer and pelvic size became limiting. At this point birth would have been difficult and dangerous and the assistance of others at the birth increasingly a necessity. We must therefore posit a level of cultural and cognitive development sufficient to support a woman in childbirth. The infant itself would be increasingly altricial and helpless at birth, with increased dependency in the postnatal period. Lactation would be an increasing metabolic load on the mother to support postnatal growth, including continued brain growth. Social and material support for the mother during the period of infant dependency would thus also become increasingly important.

Other life history parameters might also have changed at this point. As female pelvic dimensions become an increasing constraint on successful parturition, the period of growth prior to reproductive maturation lengthens. As cultural adaptation and social organization become more complex and effective in buffering people from environmental sources of mortality, life-span extends and survivorship into the postmenopausal period becomes more and more routine. Extended life-span in turn produces a multigenerational society with a greater time-depth of experience and memory. The emerging age group of postreproductive adults would also be under strong selection to continue to contribute to the survival and reproductive success of their offspring and grand-offspring.

Out of this snowball of brain growth, cultural advancement, and life history change emerged *Homo sapiens*, who not only could manage to grow the largest-brained offspring yet, but could do so at such a prodigious reproductive rate that only a hundred thousand years or less are required for the species to spread over six continents. But the story doesn't end there. Within the last fifteen thousand years, perhaps as a result of increasing population pressure, perhaps as a result of increasing cultural sophistication, another ecological shift began, a shift toward domestication of plants and animals and toward agricultural subsistence. Energy flow became even more predictable, if often seasonal, and digestible, calorie-dense weaning foods based on cooked cereals and animal milk became available. Even more metabolic energy became available for reproduction as a result, leading to an increased birth rate. With urbanization would also have come an increasing prevalence of mortality from infectious diseases, leading to an increase in infant mortality. Still, the accelerated rate of reproduction more than compensated, ushering in a new phase of human population expansion and the marginalization of hunting and gathering as a viable subsistence strategy. Still more recent advances in agricultural production and medical and public health measures have led to increased control of infectious disease mortality, the secular trend in growth and maturation, and the modern population explosion.

Our ancestors may have been limited by their ability to support brain growth in their offspring. They developed a trick for stockpiling energy to survive famines, and then were able to use the same trick to divert more energy into pregnancy and lactation. The result was an increasingly clever animal increasingly free of its own ecological constraints. The question now is whether we have become clever enough to restrain the reproductive rate that nature no longer restrains for us. We do, in fact, find ourselves in exactly the position that Malthus originally described, needing to control our fertility through personal behavior and societal policy, but not because fertility is naturally free of biological constraint. It is our cultural success, the fruit of our biological evolution, that has freed us to such an extent from the natural constraints on fecundity that most organisms contend with, constraints whose signature we can still detect in our reproductive physiology. It may well be, however, that it was an adaptation in our reproductive physiology that originally set the stage for our intellectual and cultural development. Only the future will tell whether we will survive our own reproductive success.

THE PROCESSION

A few days after the birth, and subsequent death, of Elena's twins Pippi and I were sitting in the *baraza* after returning from our daily research rounds. The rounds themselves had been short, since most of the women who were our subjects were still at Elena's *parcel* participating in the general wake that followed the death of her twins. A noise began to filter up the road that soon resolved itself into drumming and singing. Looking up the road we were surprised to see a group of thirty or more Lese and Efe women walking in two columns up the muddy tire ruts. Each of the women was draped with sweet potato vines and cassava leaves. One Lese man accompanied them, beating rhythms on a drum to accompany their loud singing. They turned off the road onto our *parcel* and began an energetic dance. The Lese man was joined by an Efe man with another drum and the two of them created ever more complicated patterns of rhythm to support the women dancers. The dancing appeared to be led by Itei, a domineering older woman leaning on a staff and wielding a stalk of cassava in her other hand. Babies jiggled on the backs of some of the women as they circled round and round in the pattern of the dance. More and more women came to join in the dance, sitting out when they were tired with their backs against the mud walls of our buildings, then jumping up to rejoin the group. Eventually general exhaustion brought the dance to a halt, at which point Itei instructed me to distribute tobacco leaves to all. As eager hands picked coals out of our fire, I asked what the occasion for such a display might be. This was nothing, I was told. Wait until tomorrow. The purpose today was simply to spread the word about the procession tomorrow. I managed to learn that it had something to do with the twins, but no details. It was for women, I was told.

Of course we were ready and eagerly waiting the next morning. Eventually the noise of drumming and singing began to come up the road from the south. Soon the procession appeared, this time apparently including every Lese woman fit enough to walk. Again the women were draped with garden vines, but today they also carried garden tools and entire plants. The children that had danced gleefully in the party the previous day were absent. But conspicuous to us in the center of the throng was Elena, with a clean cloth draped around her body in the traditional manner. Accompanied by the two male drummers, the women marched singing by our *parcel*. We grabbed our cameras and notebooks and followed along, tentatively at first, but no one seemed to mind our presence, or even to notice us. The procession continued until it reached the northern limit of the territory traditionally claimed by the local population. At this point the singing stopped. Several women slipped off the road and down the bank to the nearby stream that marked the boundary, returning with leaves full of water and mud. Others began hastily and almost haphazardly to dig holes in the ground at the side of the road for the plants they were holding. Elena stepped forward and drew out a bowl of ashes from under her cloth, adding them to the river water and mud on the ground in the center of the road. All the women quickly stooped and scraped some of the mixture onto their garden tools. Then silently, in small groups, they left.

I hadn't dared to speak while this drama was unfolding. But later, back in the *baraza,* I questioned Baudoin. After teasing me for a bit with the suggestion that I, of all people, should understand these things, he explained that it was a ritual that had to be performed because of the twins. Not, as I thought at first, because the twins had died, but because they had been born. Twins, he explained, were dangerous. They put the crops in jeopardy. No woman had been able to work in her garden since Elena's twins were born. Instead the women had to wait until Elena herself was able to join them in this ritual. The crops they had planted by the road would absorb all the danger and die. The garden tools had been cleansed. The danger was past. If this ritual had not been performed, it was likely that the crops would have failed and the survival of the Lese would have been threatened. But now life would return to normal.

As I mulled over this story I began to recognize a symmetry between this ritual ecology and the ecology of reproduction that I was in the Ituri to study. It was as if there was an economy to fertility that the Lese intuited: only so much to go around. Elena had received more than her share, and that left a shortage in the world, a shortage that would endanger the fertility of the gardens. The fertility of the gardens and the people they supported were linked, it seemed, mutually dependent. My own version of the ecological links between the fertility of the people and that of their crops was somewhat different, with the arrows of causation pointing in the opposite direction for the most part. But there was a certain satisfying quality to the symmetry, and a resonance with the linkages between human fertility and agricultural fertility recognized and ritualized by countless human cultures the world over.

The ritual significance attached to one woman's experience of giving birth to twins also underscores another important aspect of human reproduction, the way it binds the personal world to the social world. Each birth produces more than a human being; it produces new human relationships, new threads in a fabric of kinship and community. Reproduction is the crucible for the perpetuation of our cultures as well as our genes. The fate of every birth is naturally of great significance to the parents, but also to the extended family and the broader society. Our communal fate hangs on our ability to balance our individual procreative desires and our social responsibilities. This is as true for us today as for the Lese. Perhaps we need myths and rituals of our own in the twenty-first century to remind us of these connections, to impress us anew with the linkages between our own expanding population and the fate of the earth that supports us. Like Elena, we may all need to join the procession if we are to continue to live on fertile ground.

NOTES | ACKNOWLEDGMENTS | LIST OF ILLUSTRATIONS | INDEX

NOTES

PAGE

Two Births

3 The names of the Lese villagers, although authentic, have been changed.

9 Martson Bates, *The Forest and the Sea: A Look at the Economy of Nature and the Ecology of Man* (New York: Random House, 1960).

Surviving the First Cut

22 For a review of recent research on early pregnancy factor, see H. Morton, B. E. Rolfe, and A. C. Cavanaugh, "Early pregnancy factor," *Seminars in Reproductive Endocrinology* 10:72–82 (1992).

23–24 Details of oocyte maturation are well described in Paul M. Wassarman, "The mammalian ovum," in *The Physiology of Reproduction*, ed. E. Knobil, J. D. Neill, L. L. Ewing, G. S Greenwald, C. L. Markert, and D. W. Pfaff (New York: Raven Press, 1988), pp. 69–102.

25–26 Useful introductions to the behavioral effects associated with gonadal steroids can be found in E. Hampson and D. Kimura, "Sex differences and hormonal influences on cognitive function in humans," in *Behavioral Endocrinology*, ed. J. B. Becker, S. M. Breedlove, and D. Crews (Cambridge, Mass.: MIT Press, 1992), pp. 357–398, and C. S. Carter, "Hormonal influences on human sexual behavior," in *Behavioral Endocrinology*, pp. 131–142.

28–29 A general discussion of the implantation window in mammals can be found in P. J. Hogarth, *Biology of Reproduction* (Glasgow: Blackie & Son, 1978). More specific discussion of the human case can be found in R. J. Paulson, M. V. Sauer, and A. Lobo, "Embryo implantation after human in vitro fertilization: importance of endometrial receptivity," *Fertility and Sterility* 53:870–874 (1990), and A. sychoyos, "The implantation window," *Ares Serono Symposia* 4:57–62 (1993).

32 The best description of the actions of RU486 is provided by its creator, É.-E. Baulieu, "Contragestion and other clinical applications of RU486, an antiprogesterone at the receptor," *Science* 245:1351–1357 (1989).

33 To get a sense of species variation in ovarian cycles, compare the following: V. Ramirez and C. Beyer, "The ovarian cycle of the rabbit: its neuroendocrine control," in *The Physiology of Reproduction*, pp. 1873–1892; M. E. Freeman, "The ovarian cycle of the rat," in *The Physiology of Reproduction*, pp. 1893–1928; E. Knobil and J. Hotchkiss, "The menstrual cycle and its control," in *The Physiology of Reproduction*, pp. 1971–1994.

34 The North Carolina study is described in A. J. Wilcox, C. R. Weingberg, J. F. O'Conner, D. D. Baird, J. P. Schlatterer, R. E. Canfield, E. G. Armstrong, and B. C. Nisula, "Incidence of early loss of pregnancy," *New England Journal of Medicine* 319:189–194 (1988).

34 Comparative data on pregnancy loss can be found in C. L. Erhardt, "Pregnancy losses in New York City, 1960," *American Journal of Public Health* 53:1337–1357 (1963); F. E. French and J. E. Bierman, "Probabilities of fetal mortality," *Public Health Reports* 77:835–847 (1962); W. F. Taylor, "On the methodology of measuring the probability of fetal death in a prospective study," *Human Biology* 36:86–103 (1964); and S. Harlap, P. Shiono, and S. Ramcharan, "A life table of spontaneous abortions and the effects of age, parity, and other variables," in *Human Embryonic and Fetal Death*, ed. I. B. Porter and E. B. Hook (New York: Academic Press, 1980), pp. 145–164.

35 The famously titled Hertig study is described in A. T. Hertig, J. Rock, E. C. Adams, and M. C. Menkin, "Thirty-four fertilized human ova, good, bad, and indifferent, recovered from 210 women of known fertility: a study of biologic wastage in early human pregnancy," *Pediatrics* 23:202–211 (1959).

36 The Australian researcher B. E. Rolfe reported his results in "Detection of fetal wastage," *Fertility and Sterility* 5:655–660 (1982).

36–37 Two excellent exemplars of the effort to model early embryonic loss can be found in J. W. Wood, "Fecundity and natural fertility in humans," *Oxford Reviews of Reproductive Biology* 11:61–109 (1989), and C. E. Boklage, "Survival probability of human conceptions from fertilization to term," *International Journal of Fertility* 35:75–94 (1990).

37 Holman's Bangladesh study is described in his Ph.D. dissertation: Darryl J. Holman, *Total Fecundability and Fetal Loss in Rural Bangladesh* (Ann Arbor, Mich.: University Microfilms, 1996).

38–39 Data on the frequency of chromosomal defects in aborted material can be found in D. H. Carr, "Chromosomes and abortion," *Advances in Human Genetics* 2:202–257 (1971); A. Boué, J. Boué, and A. Gropp, "Cytogenetics of pregnancy wastage," *Advances in Human Genetics* 14:1–57 (1985); and D. Warburton, "Chromosomal causes of fetal death," *Clinical Obstetrics and Gynecology* 30:268–277 (1987). The evolutionary pluses and minuses of genetic recombination are discussed in G. C. Williams, *Sex and Evolution* (Princeton: Princeton University Press, 1975); J. Tooby, "Pathogens, polymorphism, and the evolution of sex," *Journal of Theoretical Biology*, 97:557–576 (1982); and J. Maynard Smith, "The evolution of recombination," in *The Evolution of Sex: An Examination of Current Ideas*, ed. R. E. Michod and B. R. Levin (Sunderland, Mass.: Sinauer, 1988).

40–41 Data on twin pregnancies and their outcomes can be found in W. F. Powers, "Twin pregnancy: complications and treatment," *Obstetrics and Gynecology* 42:795–808 (1973), and F. Pettersson, B. Smedby, and G. Lindmark, "Outcome of twin birth: review of 1636 children born in twin birth," *Acta Paediatrica Scandinavia* 64:473–479 (1976).

42 The reproductive consequences of the Nazi occupation of the Netherlands are described in Z. Stein, M. Susser, G. Saenger, and F. Morolla, *Famine and Human Development: The Dutch Hunger Winter of 1944–45* (New York: Oxford University Press, 1975), and J. Klein, Z. Stein, and M. Susser, *Conception to Birth: Epidemiology of Prenatal Development* (New York: Oxford University Press, 1989).

42–43 Haig's argument can be found in D. Haig, "Genetic conflicts in human pregnancy," *Quarterly Review of Biology* 68:495–532 (1993).

43–44 Profet's argument can be found in M. Profet, "Menstruation as a defense against pathogens transported by sperm," *Quarterly Review of Biology* 68:335–386 (1993).

45 Challenges to Profet include: J. Clarke, "The meaning of menstruation in the elimination of abnormal embryos," *Human Reproduction* 9:1204–1207 (1994); C. A. Finn, "The meaning of menstruation," *Human Reproduction* 9:1202–1203 (1994); and C. A. Finn, "Why do women menstruate? Historical and evolutionary review," *European Journal of Obstetrics and Gynecology and Reproductive Biology* 70:3–8 (1996).

45–46 Strassmann's argument can be found in B. I. Strassmann, "The evolution of endometrial cycles and menstruation," *Quarterly Review of Biology* 71:181–220 (1996), and "Energy economy in the evolution of menstruation," *Evolutionary Anthropology* 5:157–164 (1996).

46–47 Variation in mammalian ovarian cycles is described well in J. Perry, *The Ovarian Cycle of Mammals* (London: Oliver and Boyd, 1970), and H. H. Cole and P. T. Cupps, eds., *Reproduction in Domestic Animals* (New York: Academic Press, 1977).

47–48 A short review of variation in primate ovarian cycles is included in Sarah Blaffer Hrdy and Patricia L. Whitten, "Patterning of sexual activity," in *Primate Societies*, ed. B. B. Smuts, D. L. Cheney, R. M. Seyfarth, R. W. Wrangham, and T. T. Struhsaker (Chicago: University of Chicago Press, 1987), pp. 370–384. A fuller account can be found in A. F. Dixson, *Primate Sexuality* (New York: Oxford University Press, 1998).

48–49 For discussion of prehistoric human menstrual patterns, see R. V. Short, "The evolution of human reproduction," *Proceedings of the Royal Society of London (Series B)* 195:3–24 (1976); B. I. Strassmann, "Menstrual synchrony pheromones: cause for doubt," *Human Reproduction* 14:579–580 (1999); and B. I. Strassmann and J. H. Warner, "Predictors of fecundability and conception waits among the Dogon of Mali," *American Journal of Physical Anthropology* 105:167–184 (1998).

A Time to Be Born

53–54 For a comparison of childbirth in humans and other primates, see D. G. Lindburgh, "Primate obstetrics: the biology of birth," *American Journal of Primatology*, Supplement 1:193–199 (1982); W. R. Trevethan, *Human Birth: An Evolutionary Perspective* (New York: Aldine de Gruyter, 1987); and W. R. Trevethan, "The evolution of bipedalism and assisted birth," *Medical Anthropology Quarterly* 10:287–290 (1996).

54–55 The original doula study is described in R. Sosa, J. Kennel, K. Marshall, S. Robertson, and J. Urrutia, "The effect of a supportive companion on perinatal problems, length of labor, and mother-infant interaction," *New England Journal of Medicine* 303:597–600 (1980). Follow-up studies and discussion can be found in J. Kennell, M. Klaus, S. McGrath, S. Robertson, and C. Hinkley, "Continuous emotional support during labor in a US hospital: a randomized controlled trial," *Journal of the American Medical Association* 265:2197–2201 (1991); B. Chalmers and W. Wolman, "Social support in labor—a selective review." *Journal of Psychosomatic Obstetrics and Gynecology* 14:1–15 (1993); J. Zhang, J. W. Bernasko, E. Leybovich, M. Fahs, and M. C. Hatch, "Continuous labor support from labor attendant for primiparous women: a meta-analysis," *Obstetrics and Gynecology* 88:739–744 (1996); M. H. Klaus and J. H. Kennell, "The doula: an essential ingredient of childbirth rediscovered," *Acta Paediatrica* 86:1034–1036 (1997); and L. Campero, C. Garcia, C. Diaz, O. Ortiz, S. Reynoso, and A. Langer, "'Alone, I wouldn't have known what to do': a qualitative study on social support during labor and delivery in Mexico," *Social Science and Medicine* 47(3):395–403 (1998).

55–56 A fascinating description of the development of professional obstetrics in England and the role of midwives and "gossips" can be found in A. Wilson, *The Making of Man-midwifery: Childbirth in England, 1600–1770* (Cambridge, Mass.: Harvard University Press, 1995).

57–58 The pattern of fetal brain growth in humans and nonhuman primates is discussed in A. B. Holt, D. B. Cheek, E. D. Mellits, and D. E. Hill, "Brain size and the relation of the primate to the nonprimate," in *Fetal and Postnatal Growth: Hormones and Nutrition*, ed. D. B. Cheek (New York: Wiley, 1975), pp. 23–44, and T. W. Deacon, *The Symbolic Species: The Coevolution of Language and the Brain* (New York: Norton, 1997).

58–59 For data on the tight fit between maternal pelvis and fetal head, see M. M. Abitbol, M. Bowen-Ericksen, I. Castillo, and A. Pushchin, "Prediction of difficult vaginal birth and of cesarean section for cephalopelvic disproportion in early labor," *Journal of Maternal and Fetal Medicine*, 8:51–56 (1999).

59–60 An excellent review of the evolution of human bipedalism can be found in C. O. Lovejoy, "The evolution of human walking," *Scientific American*, 259:82–89 (1988).

60–61 For discussion of the evolution of the human female pelvis, see C. Berge, R. Orban-Segebarth, and P. Schmid, "Obstetrical interpretations of the australopithecine pelvic cavity," *Journal of Human Evolution*, 13:573–587 (1984); C. Berge and J. B. Kazmierczak, "Effects of size and locomotor adaptations on the hominid pelvis: evaluation of australopithecine bipedality with a new multivariate method," *Folia Primatologica*, 46:185–204 (1986).

61 Data on the physiques and skeletal dimensions of female athletes are available in M. J. Bernink, W. B. Erich, A. L. Peltenburg, M. L. Zonderland, and I. A. Huisveld, "Height, body composition, bio-

logical maturation and training in relation to socio-economic status in girl gymnasts, swimmers, and controls," *Growth*, 47:1–12 (1983); G. E. Theintz, H. Howald, Y. Allemann, and P. C. Sizonenko, "Growth and pubertal development of young female gymnasts and swimmers: a correlation with parental data," *International Journal of Sports Medicine*, 10:87–91 (1989); A. L. Claessens, F. M. Veer, V. Stijnen, J. Lefevre, H. Maes, G. Steens, and G. Beunen, "Anthropometric characteristics of outstanding male and female gymnasts," *Journal of Sports Science*, 9:53–74 (1991); K. J. Lindner, D. J. Caine, and D. P. Johns, "Withdrawal predictors among physical and performance characteristics of female competitive gymnasts," *Journal of Sports Science*, 9:259–272 (1991); and A. L. Claessens, R. M. Malina, J. Lefevre, G. Beunen, V. Stijnen, H. Maes, and F. M. Veer, "Growth and menarcheal status of elite female gymnasts," *Medicine and Science in Sports and Exercise*, 24:755–763 (1992).

62–63 Some data on prematurity, low birthweight, and infant mortality can be found in S. Shapiro, E. R. Schlesinger, and R. E. L. Nesbitt, Jr., *Infant, Perinatal, Maternal, and Childhood Mortality in the United States* (Cambridge, Mass.: Harvard University Press, 1968), and J. T. Queenan, "Fetal therapeutics: present status and future prospects," *Clinical Obstetrics and Gynecology*, 26:407–417 (1983). The question of adaptive adjustments to gestation length is raised by N. R. Peacock in "An evolutionary perspective on the patterning of maternal investment in pregnancy," *Human Nature*, 2:351–385 (1991).

63 The fascinating report of the longest human gestation is provided by L. G. Higgins, "Prolonged pregnancy (partus serotinus)," *Lancet*, 2:1154 (1954).

63–66 The original work on sheep parturition is described in G. C. Liggins, P. C. Kennedy, and L. W. Holm, "Failure of initiation of parturition after electrocoagulation of the pituitary of the fetal lamb," *American Journal of Obstetrics and Gynecology*, 98:1080–1086 (1967); G. C. Liggins, "Premature parturition after infusion of corticotrophin or cortisol into foetal lambs," *Journal of Endocrinology*, 42:323–329 (1968); and G. C. Liggins, R. J. Fairclough, S. A. Grieves, J. Z. Kendall, and B. S. Knox, "The mechanism of initiation of parturition in the ewe," *Recent Progress in Hormone Research*, 29:111–159 (1973).

66–67 Authoritative accounts of the events surrounding human parturition can be found in F. G. Cunningham, P. C. MacDonald, and N. F. Grant, *Williams' Obstetrics*, 18th ed. (Norwalk, Conn.: Appleton and Lange, 1989), and M. S. Soloff, "Endocrine control of parturition," in *Biology of the Uterus*, 2nd ed., ed. R. M. Wynn and W. P. Jollie (New York: Plenum, 1989), pp. 559–608.

67–68 For a different argument, see A.-R. Fuchs, F. Fuchs, and P. Husslein, "Oxytocin receptors and human parturition: a dual role for oxytocin in the initiation of labor," *Science*, 215:1396–1398 (1982).

68–69 The role of prostaglandins is emphasized in D. M. Strickland, S. A. Saeed, M. L. Casey, and M. D. Mitchell, "Stimulation of prostaglandin biosynthesis by urine of the human fetus may serve as a trigger for parturition," *Science*, 220:521–522 (1982).

69 The quotation is from Cunningham, MacDonald, and Grant, *Williams' Obstetrics*, p. 207.

71 Variation in gestation length and its causes is discussed in Cunningham, MacDonald, and Grant, *Williams' Obstetrics*.

71 The pattern of fetal growth is described in W. E. Brenner, D. A. Edleman, and C. H. Hendricks, "A standard of fetal growth for the United States of America," *American Journal of Obstetrics and Gynecology*, 126:555–564 (1976).

72 Maternal metabolism in pregnancy, and the concept of "accelerated starvation," are discussed in N. Freinkel, "1980 Banting Lecture: of pregnancy and progeny," *Diabetes*, 29:1023–1035 (1980), and C. J. Homko, E. Sivan, E. A. Reece, and G. Boden, "Fuel metabolism during pregnancy," *Seminars in Reproductive Endocrinology*, 17:119–125 (1999).

72–74 One useful summary of the Gambian studies carried out by the Dunn Nutrition Laboratory appears in A. M. Prentice, R. G. Whitehead, S. B. Roberts, and A. A. Paul, "Long-term energy balance in child-bearing Gambian women," *American Journal of Clinical Nutrition*, 34:2790–2799 (1981). Comparison with Western women is included in A. M. Prentice and R. G. Whitehead, "The energetics of human reproduction," *Symposia of the Zoological Society of London (Series B)*, 57:275–304 (1987).

74–75 Ramsey's experiments are described in E. M. Ramsey, G. W. Corner, Jr., and M. W. Donner, "Serial and cineradiographic visualization of maternal circulation in the primate (haemochorial) placenta," *American Journal of Obstetrics and Gynecology*, 86:213–225 (1963), and summarized in E. M. Ramsey and M. W. Donner, *Placental Vasculature and Circulation* (Philadelphia: Saunders, 1980). The effects of high-altitude hypoxia on placental function are discussed in J. A. Lichty, R. Y. Ting, P. D. Bruns, and E. Dyar, "Studies of babies born at high altitude. Part I. Relation of altitude to birth weight," *American Journal of Diseases in Childhood*, 93:666–669 (1957), and S. Zamudio, S. K. Palmer, T. Droma, E. Stamm, C. Coffin, and

L. G. Moore, "Effect of altitude on uterine artery blood flow during normal pregnancy," *Journal of Applied Physiology*, 79:7–14 (1995).

76 Comparative data on sheep and human placental glucose transfer are provided in F. H. Morriss, Jr. and R. D. H. Boyd, "Placental transport," in *The Physiology of Reproduction*, ed. E. Knobil, J. D. Neill, L. L. Ewing, G. S. Greenwald, C. L. Markert, and D. W. Pfaff (New York: Raven Press, 1988), pp. 2043–2084.

80 Discussion of the metabolic costs of the human brain can be found in E. Armstrong, "Relative brain size and metabolism in mammals," *Science*, 220:1302–1304 (1983); R. D. Martin, *Human Brain Evolution in an Ecological Context*, the 52nd James Arthur Lecture on the Evolution of the Human Brain (New York: American Museum of Natural History, 1983); Deacon, *The Symbolic Species;* and Terrence W. Deacon, "What makes the human brain different?" *Annual Review of Anthropology*, 26:337–357 (1997).

The Elixir of Life

83–84 The multiple benefits of human milk to the infant are described in D. B. Jelliffe and E. F. P. Jelliffe, *Human Milk in the Modern World* (Oxford: Oxford University Press, 1978), and J. Mestecky, C. Blair, and P. Ogra, eds., *Immunology of Milk and the Neonate* (New York: Plenum, 1991).

84 The original description of weanling diarrhea is given in J. E. Gordon, I. D. Chitkara, and J. B. Wyon, "Weanling diarrhea," *American Journal of the Medical Sciences*, March:129–161 (1963), with additional discussion in J. B. Wyon and J. E. Gordon, *The Khana Study: Population Problems in the Rural Punjab* (Cambridge, Mass.: Harvard University Press, 1971). The relationship of breast-feeding to infant infection in the United States is documented in J. Raisler, C. Alexander, and P. O'Campo, "Breast-feeding and infant illness: a dose-response relationship?" *American Journal of Public Health*, 89:25–30 (1999).

84–85 Description of mammary development can be found in I. A. Forsyth, "Mammary development," *Proceedings of the Nutrition Society*, 48:17–22 (1989); I. A. Forsyth, "The mammary gland," *Baillieres Clinics in Endocrinology and Metabolism*, 5:809–832 (1991). For a discussion of the cancer-promoting effects of estradiol, see P. T. Ellison, "Reproductive ecology and reproductive cancers," in *Hormones, Health, and Human Behavior: A Socioecological and Lifespan Perspective*, ed. C. Panter-Brick and C. M. Worthman (Cambridge: Cambridge University Press, 1999), pp. 184–209, and J. G. Lier, "Is estradiol a genotoxic mutagenic carcinogen?" *Endocrine Reviews*, 21:40–54 (2000).

85–86 A general description of the physiology of lactation can be found in H. A. Tucker, "Lactation and its hormonal control," in *The Physiology of Reproduction*, ed. E. Knobil, J. D. Neill, L. L. Ewing, G. S. Greenwald, C. L. Markert, and D. W. Pfaff (New York: Raven Press, 1988), pp. 2235–2264.

86–87 The neural circuitry involved in lactation is discussed in A. S. McNeilly, C. A. F. Robinson, M. J. Houston, and P. W. Howie, "Release of oxytocin and PRL in response to suckling," *British Medical Journal*, 286: 257–259 (1983); J. B. Wakerly, G. Clarke, and A. J. S. Summerlee, "Milk ejection and its control," in *The Physiology of Reproduction*, pp. 2283–2322; Roger V. Short, "Breast feeding," *Scientific American*, 250:35–41 (1984); and J. E. Tyson, "Neuroendorine control of lactational infertility," *Journal of Biosocial Science*, Supplement 4:23–40 (1977).

88 The effect of prolactin on insulin receptors is described in D. J. Flint, P. A. Sinnett-Smith, R. A. Clegg, and R. G. Vernon, "Role of insulin receptors in the changing metabolism of adipose tissue during pregnancy and lactation in the rat," *Biochemistry Journal*, 182:421–427 (1979); D. E. Bauman and W. B. Currie, "Partitioning of nutrients during pregnancy and lactation: a review of mechanisms involving homeostasis and homeorhesis," *Journal of Dairy Science*, 63:1514–1529 (1980); and D. J. Flint, "Role of insulin and the insulin receptor in nutrient partitioning between the mammary gland and adipose tissue," *Biochemical Society Transactions*, 13:828–829 (1985). The results of experimental manipulation of insulin levels on milk production in cows are described in D. J. Kronfeld, G. P. Mayer, J. McD. Robertson, and F. Raggi, "Depression of milk secretion during insulin administration," *Journal of Dairy Science*, 46:559–563 (1963). An overview of fat metabolism in lactation can be found in J. P. McNamara, "Role and regulation of metabolism in adipose tissue during lactation," *Journal of Nutritional Biochemistry*, 6:120–129 (1995).

89 The interaction of steroid contraception with lactation and its policy implications are discussed in M. Carballo, "The provision of contraceptive methods during lactation and support for breast-feeding: policies and practice," *International Journal of Gynaecology and Obstetrics*, Supplement 25:27–45 (1987); S. Koetsawang, "The effects of contraceptive methods on the quality and quantity of breast milk," *International Journal of Gynaecology and Obstetrics*, Supplement 25:115–127 (1987); and V. H. Laukaran, "The effects of con-

traceptive use on the initiation and duration of lactation," *International Journal of Gynaecology and Obstetrics*, Supplement 25:129–142 (1987).

89–90 Estimates of the energy requirements necessary to support lactation are provided in Food and Agriculture Organization/World Health Organization/United Nations University, "Energy and protein requirements," *World Health Organization Technical Report*, Series 724:1–206 (1985), and A. M. Prentice and A. Prentice, "Maternal energy requirements to support lactation," in *Breastfeeding, Nutrition, Infection, and Infant Growth in Developed and Emerging Countries*, ed. S. A. Atkinson, L. A. Hanson, and R. K. Chandra (St. John's, Newfoundland: ARTS Biomedical, 1990), pp. 67–86.

90 Discussion of the means by which energy can be mobilized to support lactation can be found in A. M. Prentice and R. G. Whitehead, "The energetics of human reproduction," *Symposia of the Zoological Society of London*, Series B, 57:275–304 (1987), and A. M. Prentice, "Adaptations to long-term low energy intake," in *Energy Intake and Activity*, ed. E. Pollitt and P. Amante (New York: Alan R. Liss, 1984), pp. 3–31. Panter-Brick's study of Tamang women is summarized in C. Panter-Brick, "Seasonality of energy expenditure during pregnancy and lactation for rural Nepali women," *American Journal of Clinical Nutrition*, 57:620–628 (1993).

90–91 The Gambian data are presented and discussed in various forms in Prentice and Whitehead, "The energetics of human reproduction"; R. G. Whitehead, Alison A. Paul, and M. G. M. Rowland, "Lactation in Cambridge and in the Gambia," *Topics in Paediatrics*, 2: 22–33 (1980); A. M. Prentice, R. G. Whitehead, S. B. Roberts, and A. A. Paul, "Long-term energy balance in child-bearing Gambian women," *American Journal of Clinical Nutrition*, 34:2790–2799 (1981); and S. B. Roberts, A. A. Paul, T. J. Cole, and R. G. Whitehead, "Seasonal changes in activity, birth weight, and lactational performance in rural Gambian women," *Transactions of the Royal Society of Tropical Medicine and Hygiene*, 76:667–678 (1982).

91 The supplement program is described in A. M. Prentice, T. J. Cole, F. A. Foord, W. H. Lamb, and R. G. Whitehead, "Increased birthweight after prenatal dietary supplementation of rural African women," *American Journal of Clinical Nutrition*, 46:912–925 (1987).

91–92 Data on the effects of nutritional stress on human milk production are available in A. M. Prentice, S. B. Roberts, A. Prentice, A. A. Paul, M. Watkinson, A. A. Watkinson, and R. G. Whitehead, "Dietary supplementation of lactating Gambian women. I. Effect on breast-milk volume and quality," *Human Nutrition: Clinical Nutri-*

tion, 37C:53–64 (1983), and N. F. Butte, C. Garza, J. E. Stuff, E. O'Brian Smith, and B. L. Nichols, "Effect of maternal diet and body composition on lactational performance," *American Journal of Clinical Nutrition*, 39:296–306 (1984).

92 The effects of exercise on milk production are described in C. A. Lovelady, B. Lonnerdal, and K. G. Dewey, "Lactation performance of exercising women," *American Journal of Clinical Nutrition*, 52:103–109 (1990), and K. G. Dewey, C. A Lovelady, L. A. Nommsen-Rivers, M. A. McCrory, and B. Lonnerdal, "A randomized study of the effects of aerobic exercise by lactating women on breast-milk volume and composition," *New England Journal of Medicine*, 330:449–453 (1994).

92–93 The effects of maternal diet on milk composition are described in B. Lonnerdal, "Effects of maternal dietary intake on human milk composition," *Journal of Nutrition*, 116:499–513 (1986).

93–94 The term homeorhesis is introduced in D. E. Bauman, J. H. Eisemann, and W. B. Currie, "Hormonal effects on partitioning of nutrients for tissue growth: role of growth hormone and prolactin," *Federation Proceedings*, 41:2538–2544 (1982). Exercise responses in lactating women are described in M. Altemus, P. A. Deuster, E. Gallivan, C. Sue Carter, and P. W. Gold, "Suppression of hypothalamic-pituitary-adrenal axis responses to stress in lactating women," *Journal of Clinical Endocrinology and Metabolism*, 80:2954–2959 (1995). Related metabolic responses are described in P. J. Illingworth, R. T. Jung, P. W. Howie, P. Leslie, and T. E. Isles, "Dimunition in energy expenditure during lactation," *British Medical Journal*, 292:437–441 (1986).

94–95 The concept of the maternal depletion syndrome is developed in D. B. Jelliffe and I. Maddocks, "Notes on ecological malnutrition in the New Guinea Highlands," *Clinical Pediatrics*, 3:432–438 (1964); D. B. Jelliffe, *The Assessment of the Nutritional Status of the Community* (Geneva: World Health Organization, 1966); D. P. Tracer, "Fertility-related changes in maternal body composition among the Au of Papua New Guinea," *American Journal of Physical Anthropology*, 85:393–405 (1991); A. Winkvist, K. M. Rasmussen, and J.-P. Habicht, "A new definition of maternal depletion syndrome," *American Journal of Public Health*, 82:691–694 (1992); and J. E. Miller, G. Rodriguez, and A. R. Pebley, "Lactation, seasonality, and mother's postpartum weight change in Bangladesh: an analysis of maternal depletion," *American Journal of Human Biology*, 6:511–524 (1994).

95–97 Data on the risks associated with closely spaced births are available in S. O. Rutstein, "Infant and child mortality: levels, trends, and demographic differentials," *WFS Comparative Studies*, no. 24 (London: World Fertility Survey, 1983); J. Trussell and C. Hammerslough, "A hazards-model analysis of the covariates of infant and child mortality in Sri Lanka," *Demography*, 20:1–26 (1983); J. G. Cleland and Z. Sathar, "The effect of birth-spacing on childhood mortality in Pakistan," *Population Studies*, 38:401–418 (1984); A. L. Adlakha and C. M. Suchindran, "Factors affecting infant and child mortality," *Journal of Biosocial Science*, 17:481–496 (1985); A. Palloni and S. Millman, "The effects of inter-birth intervals and breastfeeding on infant and early childhood mortality," *Population Studies*, 40:215–236 (1986); A. Palloni and M. Tienda, "The effects of breastfeeding and pace of childbearing on mortality at early ages," *Demography*, 23:31–52 (1986); J. Hobcraft, J. W. McDonald, and S. Rutstein, "Child-spacing effects on infant and early child mortality," *Population Index*, 49:585–618 (1983); and R. D. Retherford, M. K. Choe, S. Thapa, and B. B. Gubhaju, "To what extent does breastfeeding explain birth-interval effects on early childhood mortality?" *Demography*, 26:439–450 (1989). Data from the United States can be found in J. S. Rawlings, V. B. Rawlings, and J. A. Read, "Prevalence of low birth weight and preterm delivery in relation to the interval between pregnancies among white and black women," *New England Journal of Medicine*, 332:69–74 (1995).

97 A wonderful discussion of the history of wet-nursing can be found in S. Blaffer Hrdy, *Mother Nature: A History of Mothers, Infants, and Natural Selection* (New York: Pantheon, 1999).

97–100 For an excellent biography of Malthus by a distinguished modern demographer, see W. Petersen, *Malthus* (Cambridge, Mass.: Harvard University Press, 1979). For a modern edition of Malthus's famous essay with a useful introduction, see T. Malthus, *An Essay on the Principle of Population* (New York: Viking Penguin, 1970).

101 F. Lorimer, ed., *Culture and Human Fertility* (Geneva: UNESCO, 1954).

101–102 Davis and Blake's landmark paper is K. Davis and J. Blake, "Social structure and fertility: an analytic framework," *Economic Development and Culture Change*, 4:211–235 (1956).

102–103 For the development and application of Bangaarts's proximate determinant approach, see J. Bongaarts, "Intermediate fertility variables and marital fertility rates," *Population Studies*, 30:227–241 (1976); J. Bongaarts and R. G. Potter, *Fertility, Biology, and Behavior: An Analysis of the Proximate Determinants* (New York: Academic Press, 1983); and Kenneth L. Campbell and James W. Wood, "Fertility in traditional societies," in *Natural Human Fertility: Social and Biological Determinants*, ed. P. Diggory, M. Potts, and S. Teper (London: Macmillan, 1988), pp. 36–69.

103 Henry introduces the concept of "natural fertility" in Louis Henry, "Some data on natural fertility," *Eugenics Quarterly*, 8:81–91 (1961).

106 The quotations are from Henry, "Some data on natural fertility," p. 91.

106–107 Early data on postpartum amenorrhea are summarized in C. H. Peckham, "An investigation of some effects of pregnancy noted six weeks and one year after delivery," *Bulletin of the Johns Hopkins Hospital*, 54:186–207 (1934); M. Booth, "The time of reappearance and the character of the menstrual cycle following gestation," *Yale Journal of Biology and Medicine*, 1935:215–216 (1935); P. M. Lass, J. Smelser, and P. Kurzrok, "Studies relating to time of human ovulation. III. During lactation," *Endocrinology*, 23:39–43 (1938); P. Topkins, "The histologic reappearance of the endometrium during lactation amenorrhea and its relationship to ovarian function," *American Journal of Obstetrics and Gynecology*, 45:48–58 (1943); and I. C. Udesky, "Ovulation in lactating women," *American Journal of Obstetrics and Gynecology*, 59:843–851 (1950).

107 Early data on postpartum amenorrhea from non-Western populations appear in W. Martin, D. Morley, and M. Woodland, "Intervals between births in a Nigerian village," *Journal of Tropical Pediatrics*, 10:82–85 (1964); R. G. Potter, M. L. New, J. B. Wyon, and J. E. Gordon, "A fertility differential in eleven Punjab villages," *Milbank Memorial Fund Quarterly*, 43:185–201 (1965); M. Bonte and H. van Balen, "Prolonged lactation and family spacing in Rwanda," *Journal of Biosocial Science*, 1:97–100 (1969); A. K. Jain, C. Hsu, R. Freedman, and M. C. Chang, "Demographic aspects of lactation and postpartum amenorrhea," *Demography*, 7:255–271 (1970); L. C. Chen, S. Ahmed, M. Gesche, and W. H. Mosely, "A prospective study of birth interval dynamics in rural Bangladesh," *Population Studies*, 28:277–297 (1974); and J. K. Van Ginneken, "The chance of conception during lactation," *Journal of Biosocial Science*, Supplement 4:41–54 (1977).

107–108 Early data on the effects of "full" and "partial" breastfeeding can be found in A. Sharman, "Menstruation after childbirth," *Journal of Obstetrics and Gynaecology of the British Empire*, 58::440–445 (1951), and T. McKeown and J. R. Gibson, "A note on menstruation and conception during lactation," *Journal of Obstetrics and Gynaecology of the British Empire*, 61:824–829 (1954).

108 Arguments implicating prolactin in the suppression of menstruation are found in C. Robyn, P. Delvoye, C. Van Exter, M. Vekemans, A Caufriez, P. de Nayer, J. Delogne-Desnoeck, and M. L'Hermite, "Physiological and pharmacological factors influencing prolactin secretion and their relation to human reproduction," in *Prolactin and Human Reproduction*, ed. P. G. Crosignani and C. Robyn (New York: Academic Press, 1976), pp. 71–96, and P. Delvoye, J. Desnoeck-Delogne, and C. Robyn, "Serum-prolactin in long-lasting amenorrhea," *Lancet*, 2:228 (1976). Tyson's studies of prolactin dynamics are summarized in J. E. Tyson, "Neuroendocrine control of lactational infertility," *Journal of Biosocial Science*, Supplement 4:23–40 (1977).

109–110 The studies of nursing frequency and prolactin in Zaire are summarized in P. Delvoye, M. Demaegd-Delogne, and C. Robyn, "The influence of the frequency of nursing and of previous lactation experience on serum prolactin in lactating mothers," *Journal of Biosocial Science*, 9:447–451 (1977), and P. Delvoye, M. Demaegd, Uwayitu-Nyampeta, and C. Robyn, "Serum prolactin, gonadotropins, and estradiol in menstruating and amenorrheic mothers during two years' lactation," *American Journal of Obstetrics and Gynecology*, 130:635–639 (1978).

110–111 Konner and Worthman's studies of nursing frequency among the !Kung are described in M. Konner and C. Worthman, "Nursing frequency, gonadal function, and birth spacing among !Kung hunter-gatherers," *Science*, 207:788–791 (1980).

112–114 The original results of the Edinburgh studies are presented in a series of papers all bearing the primary title "Fertility after childbirth": P. W. Howie, A. S. McNeilly, M. J. Houston, A. Cook, and H. Boyle, "Fertility after childbirth: infant feeding patterns, basal PRL levels, and post-partum ovulation," *Clinical Endocrinology*, 17:315–322 (1982); P. W. Howie, A. S. McNeilly, M. J. Houston, A. Cook, and H. Boyle, "Fertility after childbirth: post-partum ovulation and menstruation in bottle and breast feeding mothers," *Clinical Endocrinology*, 17:323–332 (1982); A. S. McNeilly, P. W. Howie, M. J. Houston, A. Cook, and H. Boyle, "Fertility after childbirth: adequacy of post-partum luteal phases," *Clinical Endocrinology*, 17:609–615 (1982); A. S. McNeilly, A. Glasier, P. W. Howie, M. J. Houston, A. Cook, and H. Boyle, "Fertility after childbirth: pregnancy associated with breast feeding," *Clinical Endocrinology*, 18:167–173 (1983); and A. Glasier, A. S. McNeilly, and P. W. Howie, "Fertility after childbirth: changes in serum gonadotrophin levels in bottle and breast feeding women," *Clinical*

Endocrinology, 19:493–501 (1983). A useful overview of these studies and their significance appears in P. W. Howie and A. S. McNeilly, "Effect of breast feeding patterns on human birth intervals," *Journal of Reproduction and Fertility*, 65:545–557 (1982).

115 Short's argument can be found in Roger V. Short, "The biological basis for the contraceptive effects of breast feeding," *International Journal of Gynaecology and Obstetrics*, Supplement 25:207–217 (1987).

115–116 The statement of the Bellagio consensus appears in K. I. Kennedy, R. Rivera, and A. S. McNeilly, "Consensus statement on the use of breastfeeding as a family planning method," *Contraception*, 39:477–496 (1989).

116 Although he did not articulate it in the terms presented here, the metabolic load hypothesis owes much to Peter G. Lunn, "Lactation and other metabolic loads affecting human reproduction," *Annals of the New York Academy of Sciences*, 709:77–85 (1994).

116–117 Changing views on the physiology of lactational amenorrhea are reviewed in Alan S. McNeilly, Clem C. K. Tay, and Anna Glasier, "Physiological mechanisms underlying lactational amenorrhea," *Annals of the New York Academy of Sciences*, 709:145–155 (1994), and Alan. S. McNeilly, "Breastfeeding and the baby," in *Human Reproductive Decisions: Biological and Social Perspectives*, ed. R. I. M. Dunbar (London: Macmillan, 1995), pp. 9–21. An argument for the continued relevance of prolactin as an index of nursing intensity appears in V. J. Vitzthum, "Comparative study of breastfeeding structure and its relation to human reproductive ecology," *Yearbook of Physical Anthropology*, 37:307–349 (1994). New data on nursing patterns and prolactin responses that challenge the previous paradigm can be found in C. C. Tay, A. F. Glasier, and A. S. McNeilly, "Twenty-four hour patterns of prolactin secretion during lactation and the relationship to suckling and the resumption of fertility in breast-feeding women," *Human Reproduction*, 11:950–955 (1996), and J. F. Stallings, C. M. Worthman, C. Panter-Brick, and R. J. Coates, "Prolactin response to suckling and maintenance of postpartum amenorrhea among intensively breastfeeding Nepali women," *Endocrine Research* 22:1–28 (1996).

117 The Bangladeshi and Australian studies are described in S. L. Huffman, A. Chowdhury, H. Allen, and L. Nahar, "Suckling patterns and post-partum amenorrhea in Bangladesh," *Journal of Biosocial Science*, 19:171–179 (1987), and Patricia R. Lewis, James B. Brown, Marilyn B. Renfree, and Roger V. Short, "The resumption of ovulation and menstruation in a well-nourished population of women

breastfeeding for an extended period of time," *Fertility and Sterility*, 55:529–536 (1991).

118–119 Data for the Yolgnu, Amele, and Toba are from Janet W. Rich, "Patterns of breast feeding and lactational infertility: a comparison of the Yolngu with the !Kung San and La Leche League," B.A. Honors Thesis, Department of Anthropology, Harvard University, 1984; C. M. Worthman, C. L. Jenkins, J. F. Stallings, and D. Lai, "Attenuation of nursing-related ovarian suppression and high fertility in well-nourished, intensively breast-feeding Amele women of lowland Papua New Guinea," *Journal of Biosocial Science*, 25:425–443 (1993); and C. R. Valeggia and P. T. Ellison, "Nursing patterns, maternal energetics, and postpartum fertility among Tobas of Formosa, Argentina," *American Journal of Physical Anthropology*, Supplement 26:222 (1998).

119–120 The Gambian data are presented in P. G. Lunn, A. M. Prentice, S. Austin, and R. G. Whitehead, "Influence of maternal diet on plasma-prolactin levels during lactation," *Lancet*, 1:623–625 (1980), and P. G. Lunn, S. Austin, A. M. Prentice, and R. G. Whitehead, "The effect of improved nutrition on plasma prolactin concentrations and postpartum infertility in lactating Gambian women," *American Journal of Clinical Nutrition*, 39:227–235 (1984).

121 Two recent reviews of the literature on nursing behavior and amenorrhea are Vitzthum, "Comparative study of breastfeeding structure and its relation to human reproductive ecology" and P. T. Ellison, "Breastfeeding, fertility, and maternal condition," in *Breastfeeding: Biocultural Perspectives*, ed. P. Stuart-Macadam and K. A. Dettwyler (New York: Aldine de Gruyter, 1995), pp. 305–345.

121 Evidence that maternal constraints affect nursing patterns can be found in C. Panter-Brick, "Motherhood and subsistence work—the Tamang of rural Nepal," *Human Ecology*, 17:205–228 (1989); C. Panter-Brick, "Working mothers in rural Nepal," in *The Anthropology of Breast-Feeding*, ed. V. Maher (Oxford: Berg, 1992), pp. 133–150; and Virginia J. Vitzthum, "Nursing behavior and its relation to duration of post-partum amenorrhoea in an Andean community," *Journal of Biosocial Science*, 21:145–160 (1989). Evidence that maternal differences affect nursing patterns in primates can be found in R. A. Hinde, *Biological Bases of Human Social Behavior* (New York: McGraw-Hill, 1974); J. Altmann, *Baboon Mothers and Infants* (Cambridge, Mass.: Harvard University Press, 1980); Nancy A. Nicholson, "Infants, mothers, and other females," in *Primate Societies*, ed. B. B. Smuts, D. L. Cheney, R. M. Seyfarth, R. W. Wrang-

ham, and T. T. Struhsaker (Chicago: University of Chicago Press, 1987), pp. 330–342; M. Gomendio, "Differences in fertility and suckling patterns between primiparous and multiparous rhesus mothers *(Macaca mulatta)*," *Journal of Reproduction and Fertility*, 87:1–14 (1990); and M. Gomendio, "The influence of maternal rank and infant sex on maternal investment in rhesus macaques: birth sex ratios, inter-birth intervals and suckling patterns," *Behavioral Ecology and Sociobiology*, 27:365–375 (1990).

124–125 The Frisch-Bongaarts debate makes fascinating reading: R. E. Frisch, "Population, food intake, and fertility," *Science*, 199:22–30 (1978); J. Bongaarts, "Does malnutrition affect fecundity? A summary of evidence," *Science*, 208:564–569 (1980); R. E. Frisch, "Malnutrition and fertility," *Science*, 215:1272–1273 (1982); and J. Bongaarts, "Malnutrition and fertility," *Science*, 215:1273–1274 (1982).

125 Wood's argument can be found in J. W. Wood, "Maternal nutrition and reproduction: why demographers and physiologists disagree about a fundamental relationship," *Annals of the New York Academy of Sciences*, 709:101–116 (1994), and J. W. Wood, *Dynamics of Human Reproduction: Biology, Biometry, Demography* (New York: Aldine de Gruyter, 1994).

Why Grow Up?

129–130 The view that human maturation is delayed to allow for extended social and cognitive development is championed by B. Bogin in *Patterns of Human Growth* (Cambridge: Cambridge University Press, 1988). Evidence of the dramatic variability in human maturation is summarized in P. B. Eveleth and J. M. Tanner, *Worldwide Variation in Human Growth*, 2nd ed. (Cambridge: Cambridge University Press, 1990).

130 Useful reviews of the patterns of human growth are provided in I. Valadian and D. Porter, *Physical Growth and Maturity from Conception to Maturity* (Boston: Little, Brown, 1977), and J. M. Tanner, *Fetus into Man: Physical Growth from Conception to Maturity* (Cambridge, Mass.: Harvard University Press, 1990).

131–132 The basic argument for the relevance of universal growth standards is developed in J.-P. Habicht, R. Martorell, C. Yarbough, R. M. Malina, and R. E. Klein, "Height and weight standards for preschool children: how relevant are ethnic differences in growth potential?" *Lancet*, 1:611–615 (1974). The appropriate use of growth standards is discussed in J. M. Tanner, "Use and abuse of growth standards," in *Human Growth,* 2nd ed., ed. F. Falkner and J. M. Tanner (New York: Plenum, 1986).

134–135 For data on heritability of physical and developmental variables, see L. L. Cavalli-Sforza and W. F. Bodmer, *The Genetics of Human Population* (San Francisco: W. H. Freeman, 1971); S. Fischbein and T. Nordqvist, "Profile comparisons of physical growth for monozygotic and dizygotic twin pairs," *Annals of Human Biology,* 5:321–328 (1978); J. C. Sharma, "The genetic contribution to pubertal growth and development studied by longitudinal growth data on twins," *Annals of Human Biology,* 10:163–171 (1983); S. Fischbein, "Onset of puberty in MZ and DZ twins," *Acta Geneticae Medicae et Gemellologiae,* 26:151–158 (1977); and B. O. Ljung, S. Fischbein, and G. Lindgren, "A comparison of growth in twins and singleton controls of matched age followed longitudinally from 10 to 18 years," *Annals of Human Biology,* 4:405–415 (1977).

136 Ethnic differences in growth during childhood and adolescence are documented in A. R. Frisancho, K. Guire, W. Babler, G. Borkan, and A. Way, "Nutritional influence on childhood development and genetic control of adolescent growth of Quechuas and mestizos from the Peruvian lowlands," *American Journal of Physical Anthropology,* 52:367–375 (1980).

137 Descriptions of the use of hand-wrist x-rays in assessing development can be found in Tanner, *Fetus into Man,* and J. M. Tanner, R. H. Whitehouse, N. Cameron, W. A. Marshall, M. J. Healy, and H. Goldstein, *Assessment of Skeletal Maturity,* 2nd ed. (London: Academic Press, 1983).

137–138 Descriptions of secondary sexual characteristic development can be found in J. M. Tanner, *Growth at Adolescence,* 2nd ed. (Oxford: Blackwell, 1962); N. G. Norgan, "Body composition," in *The Cambridge Encyclopedia of Human Growth and Development,* ed. S. J. Ulijaszek, F. E. Johnston, and M. A. Preece (Cambridge: Cambridge University Press, 1998), pp. 221–215; P. T. Ellison, "Sexual maturation," in *The Cambridge Encyclopedia of Human Growth and Development,* pp. 227–229; P. T. Ellison, "Skeletal growth, fatness, and menarcheal age: a comparison of two hypotheses," *Human Biology,* 54:269–281 (1982); and M. L. Moerman, "Growth of the birth canal in adolescent girls," *American Journal of Obstetrics and Gynecology,* 143:528–532 (1982).

138 The sequence of pubertal development is described in Tanner, *Growth at Adolescence;* W. A. Marshall and J. M. Tanner, "Variation

in the pattern of pubertal changes in boys," *Archives of Diseases in Childhood*, 45:13–23 (1970); and W. A. Marshall and J. M. Tanner, "Puberty," in Tanner and Falkner, *Human Growth*, pp. 171–210.

138–139 Disorders of pubertal development are described in D. M. Styne and M. M. Grumbach, "Disorders of puberty in the male and female," in *Reproductive Endocrinology: Physiology, Pathophysiology, and Clinical Management*, 3rd ed., ed. S. S. C. Yen and R. B. Jaffe (Philadelphia: Saunders, 1991), pp. 313–384. Boas's studies are summarized in Franz Boas, "Studies in growth," *Human Biology*, 4:307–350 (1932); "Studies in growth: II," *Human Biology*, 5:429–444 (1933); "Studies in growth: III," *Human Biology*, 7:303–318 (1935). The modern method of assessing skeletal age is described in J. M. Tanner, R. H. Whitehouse, W. A. Marshall, M. J. R. Healy, and H. Goldstein, *Assessment of Skeletal Maturity and Prediction of Adult Height: TW2 Method* (New York: Academic, 1975).

139–140 The endocrinology of growth is summarized in L. E. Underwood and J. J. Van Wyk, "Hormones in normal and aberrant growth," in *Textbook of Endocrinology*, 6th ed., ed. R. H. Williams (Philadelphia: Saunders, 1981), pp. 1149–1191. The direct effect of estrogens is actually to slow mineral resorption from bones and thus to contribute to positive mineral balance in the bones; see S. R. Cummings, J. L. Kelsey, M. C. Nevitt, and K. J. O'Dowd, "Epidemiology of osteoporosis and osteoporotic fractures," *Epidemiologic Reviews*, 7:178–208 (1985), and E. F. Eriksen, D. S. Colvard, N. J. Berg , M. L. Graham, K. G. Mann, T. C. Spelsberg, and L. Riggs, "Evidence of estrogen receptors in normal human osteoblast-like cells," *Science*, 241:84–89 (1988). The case of a man without effective estrogen receptors was first described in E. P. Smith, J. Boyd, G. R. Frank, H. Takahashi, R. M. Cohen, B. Specker, T. C. Williams, D. B. Lubahn, and K. S. Korach, "Estrogen resistance caused by a mutation in the estrogen-receptor gene in a man," *New England Journal of Medicine*, 331:1056–1061 (1994), and discussed in D. Federman, "Life without estrogen," *New England Journal of Medicine*, 331:1088–1089 (1994).

140 The two-cell process of ovarian estrogen production is described in K. J. Ryan, F. Petro, and J. Kaiser, "Steroid formation by isolated and recombined ovarian granulosa and theca cells," *Journal of Endocrinology and Metabolism*, 28:355–358 (1968). The high androgen/estrogen ratios of adolescent menstrual cycles are described in D. Apter, L. Viinkka, and R. Vihko, "Hormonal pattern of adolescent menstrual cycles," *Journal of Clinical Endocrinology and Metabolism*,

47:944–954 (1978), and R. Vihko and D. Apter, "Endocrine characteristics of adolescent menstrual cycles: impact of early menarche," *Steroid Biochemistry*, 20:231–236 (1984).

140–141 The effect of congenital adrenal hyperplasia on pubertal development is documented in I. A. Hughes and G. F. Read, "Menarche and subsequent ovarian function in girls with congenital adrenal hyperplasia," *Hormone Research*, 16:100–106 (1982), and Robert B. Jaffe, "Disorders of sexual development," in Yen and Jaffe, *Reproductive Endocrinology*, pp. 480–510.

141 Support for the hypothalamic theory of puberty can be found in H. E. Kulin, M. M. Grumbach, and S. L. Kaplan, "Changing sensitivity of the pubertal gonadal hypothalamic feedback mechanism in man," *Science*, 166:1012–1013 (1969); E. Knobil, T. M. Plant, L. Wildt, P. E. Belchetz, and G. Marshall, "Control of the rhesus monkey menstrual cycle: permissive role of hypothalamic gonadotropin-releasing hormone," *Science*, 207:1371–1372 (1980); and W. F. Crowley, M. Filicori, D. I. Spratt, and N. F. Santoro, "The physiology of gonadotropin-releasing-hormone (GnRH) secretion in men and women," *Recent Progress in Hormone Research*, 41:473–531 (1985).

142 The gonadostat concept is presented in M. M. Grumbach, J. C. Roth, S. L. Kaplan, and R. P. Kelch, "Hypothalamic-pituitary regulation of puberty in man: evidence and concepts derived from clinical research," in *Control of the Onset of Puberty*, ed. M. M. Grumbach, G. D. Grave, and F. E. Fayer (New York: Wiley, 1974), pp. 115–166.

142 The pituitary drive hypothesis is described in T. M. Plant, "Puberty in primates," in *The Physiology of Reproduction*, ed. E. Knobil, J. D. Neill, L. L. Ewing, G. S Greenwald, C. L. Markert, and D. W. Pfaff (New York: Raven Press, 1988), pp. 1763–1788.

143–144 For the potential role of adrenal androgens in resetting hypothalamic sensitivity, see P. K. Siiteri, "Review of studies on estrogen biosynthesis in the human," *Cancer Research*, 42:3269s–3272s (1982), and P. K. Siiteri, "Obesity and peripheral estrogen synthesis," in *Adipose Tissue and Reproduction*, ed. R. E. Frisch (Basel: Karger, 1990), pp. 70–84.

144–145 The effects of adrenal androgens on normal and abnormal pubertal development are explored in A. R. Genazanni, V. DeLeo, P. Inaudi, and P. Kicovic, "Effects of DHAS treatment on the hypothalamus-pituitary-gonadal axis in delayed adrenarche," in *Adrenal Androgens*, ed. A. R. Genazanni, J. H. Thijssen, and P. K. Siiteri (New York: Raven Press, 1980), pp. 315–334; S. Korth-Schutz, L. S. Levine, and M. I. New, "Serum androgens in normal prepubertal and pubertal children and in children with precocious adrenarche," *Journal of Clinical Endocrinology and Metabolism*, 42:117–124 (1976); and P. Saenger and R. O. Reiter, "Editorial: Premature adrenarche: A normal variant of puberty?" *Journal of Clinical Endocrinology and Metabolism*, 74:236–238 (1992).

145–146 Frisch and Revelle's comparative study of caloric intake and the timing of adolescent growth is presented in R. E. Frisch and R. Revelle, "Variation in body weights and the age of the adolescent growth spurt among Latin American and Asian populations, in relation to calorie supplies," *Human Biology*, 41:185–212 (1969).

147 The original presentation of the critical weight hypothesis is in R. E. Frisch and R. Revelle, "Height and weight at menarche and a hypothesis of critical body weights and adolescent events," *Science*, 169:397–399 (1970). The extension of the hypothesis to cover the secular trend is presented in R. E. Frisch, "Weight at menarche: Similarity for well-nourished and undernourished girls at differing ages, and evidence for historical constancy," *Pediatrics*, 50:445–450 (1972).

148 The original data for the critical weight hypothesis are plotted in R. E. Frisch and R. Revelle, "Height and weight at menarche and a hypothesis of menarche," *Archives of Disease in Childhood*, 46:695–701 (1971).

149–152 The saga of the Frisch hypothesis, the major criticisms, responses, and revisions, can be traced through the following references: F. E. Johnston, R. M. Malina, and M. A. Galbraith, "Height, weight, and age at menarche and the 'critical weight' hypothesis," *Science*, 174:1148 (1971); R. E. Frisch, R. Revelle, and S. Cook, "Reply to Johnston, Malina, and Galbraith," *Science*, 174:1148–1149 (1971); R. E. Frisch, R. Revelle, and S. Cook, "Components of weight at menarche and the initiation of the adolescent growth spurt in girls: estimated total water, lean body weight, and fat," *Human Biology*, 45:469–483 (1973); F. E. Johnston, A. F. Rohe, L. M. Schell, and H. N. B. Wettenhall, "Critical weight at menarche: critique of a hypothesis," *American Journal of Diseases in Childhood*, 129:19–23 (1975); J. Trussell, "Menarche and fatness: reexamination of the critical body composition hypothesis," *Science*, 200:1506–1509 (1978); R. E. Frisch, "Reply to Trussell," *Science* 200:1509–1513 (1978); J. Reeves, "Estimating fatness," *Science*, 240:881 (1979); Rose E. Frisch and Janet W. McArthur, "Menstrual cycles: fatness as a determinant of minimum weight for height necessary for their onset and maintenance," *Science*, 185:949–951 (1974); W. Z. Billewicz, H. M. Fellowes, and C. A. Hytten, "Comments on the critical metabolic mass and the age of menarche," *Annals of Human Biology*, 3:51–59 (1976); R. E. Frisch, "Letter to the editor," *Annals of Human Biology*, 3:489–492 (1976); N. Cameron, "Weight and skinfold variation at menarche and the critical body weight hypothesis," *Annals of Human Biology*, 3:279–282 (1976); and P. T. Ellison, "Threshold hypotheses, developmental age, and menstrual function," *American Journal of Physical Anthropology*, 54:337–340 (1981).

153 The evolutionary rationale for the minimum fatness hypothesis is first advanced in Frisch, "Demographic implications of the biological determinants of female fecundity," *Social Biology*, 22:17–22 (1975), and reiterated in R. E. Frisch, "Population, food intake, and fertility," *Science*, 199:22–30 (1978).

154–155 Hill and Hurtado present their arguments in Kim Hill and A. Magdalena Hurtado, *Ache Life History: The Ecology and Demography of a Foraging People* (New York: Aldine de Gruyter, 1996). Charnov's approach to life history theory is summarized in Eric L. Charnov, *Life History Invariants* (Oxford: Oxford University Press, 1993).

156 Declining height variance at menarche is described in Ellison, "Threshold hypotheses, developmental age, and menstrual function." The convergence in height of early and late maturers is demonstrated in F. K. Shuttleworth, "Sexual maturation and the physical growth of girls age six to nineteen," *Monographs of the Society for Research in Child Development*, 2 (1937), 3 (1938); and Frisch and Revelle, "Height and weight at menarche and a hypothesis of critical body weights and adolescent events."

157 On predicting menarcheal age, see R. E. Frisch, "A method of prediction of age of menarche from height and weight at ages 9 through 13 years," *Pediatrics*, 53:384–390 (1974); P. T. Ellison, "Prediction of age at menarche from annual height increments," *American Journal of Physical Anthropology*, 56:71–75 (1981); and P. T. Ellison, "Skeletal growth, fatness, and menarcheal age: a comparison of two hypotheses," *Human Biology*, 54:269–281 (1982).

157 For records of precocious motherhood, see F. Labrie, "Glycoprotein hormones: gonadotropins and thyrotropin," in *Hormones: From Molecules to Disease*, ed. É.-É . Baulieau and P. A. Kelly (New York: Chapman and Hall, 1990), pp. 255–275, and E. Escomel's article in *La Presse Médicale*, 47:744, 875 (1939).

157–158 On pelvic growth and menarche, see Ellison, "Skeletal growth, fatness, and menarcheal age," and P. T. Ellison, "Human ovarian function and reproductive ecology: new hypotheses," *American Anthropologist*, 92:933–952 (1990).

158–159 The evidence for pelvic reshaping appears in Moerman, "Growth of the birth canal in adolescent girls."

162 For the development of sex differences in body composition and strength, see A. R. Frisancho, *Anthropometric Standards for the Assessment of Growth and Nutritional Status* (Ann Arbor: University of Michigan Press, 1990), and R. D. Tuddenham and M. M. Snyder, *Physical Growth of California Boys and Girls from Birth to Eighteen Years* (Berkeley: University of California Press, 1954).

Balancing Act

168–169 Studies of mate preference are summarized in David Buss, "Sex differences in human mate preferences: evolutionary hypotheses tested in 37 cultures," *Behavioral and Brain Sciences*, 12:1–49 (1989). For symmetry and mate choice in humans, see J. E. Scheib, S. W. Gangestad, and R. Thornhill, "Facial attractiveness, symmetry, and cues of good genes," *Proceedings of the Royal Society of London*, Series B, 266:1913–1917 (1999).

170 Data on male and female body composition are available in A. R. Frisancho, *Anthropometric Standards for the Assessment of Growth and Nutritional Status* (Ann Arbor: University of Michigan Press, 1990).

170–171 For discussion of starvation physiology, see A. Keys, J. Brozek, A. Henschel, and O. Mickelsen, *The Biology of Human Starvation* (Minneapolis: University of Minnesota Press, 1950), and V. R. Young and N. S. Scrimshaw, "The physiology of starvation," *Scientific American*, October:45–51 (1971).

172 The thrifty gene hypothesis was first advanced in J. V. Neel, "Diabetes mellitus: 'thrifty gene' rendered detrimental by 'progress'?" *American Journal of Human Genetics*, 14:353–362 (1962). A contemporary view can be found in S. T. McGarvey, "The thrifty gene concept and adiposity studies in biological anthropology," *Journal of the Polynesian Society*, 103:29–42 (1994).

172–173 For more recent presentations of the Frisch hypothesis, see R. E. Frisch, "Body fat, puberty, and fertility," *Biological Reviews*, 59:161–188 (1984), and R. E. Frisch "Body fat, menarche, and fertility," in *Encyclopedia of Human Nutrition* (London: Academic Press, 1998), pp. 777–785.

176 For postpartum luteal function, see A. S. McNeilly, P. W. Howie,

M. J. Houston, A. Cook, and H. Boyle, "Fertility after childbirth: adequacy of post-partum luteal phases," *Clinical Endocrinology*, 17:609–615 (1982).

176–177 The continuum of ovarian function is described in P. T. Ellison, "Human ovarian function and reproductive ecology: new hypotheses," *American Anthropologist*, 92:933–952 (1990); P. T. Ellison, "Understanding natural variation in human ovarian function," in *Human Reproductive Decisions: Biological and Social Perspectives*, ed. R. I. M. Dunbar (London: Macmillan, 1995), pp. 22–51; and P. T. Ellison, C. Panter-Brick, S. F. Lipson, and M. T. O'Rourke, "The ecological context of human ovarian function," *Human Reproduction*, 8:2248–2258 (1993).

177 For follicular growth, see S. Lenz, "Ultrasonic study of follicular maturation, ovulation, and development of corpus luteum during normal menstrual cycles," *Acta Obstetrica Gynecologica Scandinavia*, 64:15–19 (1985).

177–178 The relationship of follicular development to luteal function is discussed in G. S. DiZerega and G. D. Hodgen, "Luteal phase dysfunction infertility: a sequel to aberrant folliculogenesis," *Fertility and Sterility*, 35:489–499 (1981). Luteal cell types are discussed in G. S. Jones, "Corpus luteum: composition and function," *Fertility and Sterility*, 54:21–26 (1990).

178 For variation in cycle phase lengths, see E. A. Lenton, B.-M. Landgren, L. Sexton, and R. Harper, "Normal variation in the length of the follicular phase of the menstrual cycle: effect of chronological age," *British Journal of Obstetrics and Gynaecology*, 91:681–684 (1984), and E. A. Lenton, B.-M. Landgren, and L. Sexton, "Normal variation in the length of the luteal phase of the menstrual cycle: identification of the short luteal phase," *British Journal of Obstetrics and Gynaecology*, 91:685–689 (1984).

178–179 On anovulatory cycles, see A. N. Poindexter and M. B. Ritte, "Anovulatory uterine bleeding," *Comprehensive Therapy*, 9:65–71 (1983).

179–180 The study of estradiol levels in conception is reported in S. F. Lipson and P. T. Ellison, "Comparison of salivary steroid profiles in naturally occurring conception and non-conception cycles," *Human Reproduction*, 11:2090–2096 (1996).

180 For the significance of progesterone variation, see E. A. Lenton, R. Sulaiman, O. Sobowale, and I. D. Cooke, "The human menstrual cycle: plasma concentrations of prolactin, LH, FSH, oestradiol, and progesterone in conceiving and non-conceiving women," *Journal of Reproduction and Fertility*, 65:131–139 (1982); D. R. Stewart, J. W.

Overstreet, S. T. Nakajima, and B. L. Lasley, "Enhanced ovarian steroid secretion before implantation in early human pregnancy," *Journal of Clinical Endocrinology and Metabolism*, 76:1470–1476 (1993); M. J. McNeely and M. R. Soules, "The diagnosis of luteal phase deficiency: a critical review," *Fertility and Sterility*, 50:1–15 (1988); M. R. Soules, "Luteal-phase deficiency: the most common abnormality of the menstrual cycle?" in *The Menstrual Cycle and Its Disorders*, ed. K. M. Pirke and U. Schweiger (Berlin: Springer-Verlag, 1989), pp. 97–109; and D. R. Meldrum, "Female reproductive aging—ovarian and uterine factors," *Fertility and Sterility*, 59:1–5 (1993).

180–181 On irregular menstruation and fertility, see H. A. Kolstad, J. P. Bonde, N. H. Hjøllund, T. K. Jensen, T. B. Henriksen, E. Ernst, A. Giwercman, N. E. Skakkebæk, and J. Olsen, "Menstrual cycle pattern and fertility: a prospective follow-up study of pregnancy and early embryonal loss in 295 couples who were planning their first pregnancy," *Fertility and Sterility*, 71:490–496 (1999).

181–182 For a sampling of the extensive literature on exercise and ovarian function, see C. B. Feicht, T. S. Johnson, B. J. Martin, K. E. Sparks, and W. W. Wagner, "Secondary amenorrhoea in athletes," *Lancet*, 26:1145–1146 (1978); D. C. Cumming, G. D. Wheeler, and V. J. Harber, "Physical activity, nutrition, and reproduction," *Annals of the New York Academy of Sciences*, 709:55–76 (1994); L. Rosetta, "Female reproductive dysfunction and intense physical training," *Oxford Reviews of Reproductive Biology*, 15:113–141 (1993); D. K. Wakat, K. A. Sweeney, and A. D. Rogol, "Reproductive system function in cross country runners," *Medicine and Science in Sports and Exercise*, 14:263–269 (1982); C. F. Sanborn, B. H. Albrecht, and W. W. Wagner, "Athletic amenorrhea: lack of association with body fat," *Medicine and Science in Sports and Exercise*, 19:207–212 (1987); M. Shangold, R. Freeman, B. Thyssen, and M. Gatz, "The relationship between long-distance running, plasma progesterone, and luteal phase length," *Fertility and Sterility*, 31:699–702 (1979); J. B. Russell, D. Mitchell, P. I. Musey, and D. C. Collins, "The relationship of exercise to anovulatory cycles in female athletes: hormonal and physical characteristics," *Obstetrics and Gynecology*, 63:452–456 (1984); and R. M. Malina, R. C. Ryan, and C. M. Bonci, "Age at menarche in athletes and their mothers and sisters," *Annals of Human Biology*, 21:417–422 (1994).

182–183 The study by Bullen and her colleagues is described in B. A. Bullen, G. S. Skrinar, I. Z. Beitins, G. von Mering, B. A. Turnbull, and J. W. McArthur, "Induction of menstrual disorders by strenuous exercise in untrained women," *New England Journal of Medicine*, 312:1349–1353 (1985).

183 For studies of recreational runners, see P. T. Ellison and C. Lager, "Exercise-induced menstrual disorders," *New England Journal of Medicine*, 313:825–826 (1985); P. T. Ellison and C. Lager, "Moderate recreational running is associated with lowered salivary progesterone profiles in women," *American Journal of Obstetrics and Gynecology*, 154:1000–1003 (1986); R. E. Bledsoe, M. T. O'Rourke, and P. T. Ellison, "Characteristics of progesterone profiles of recreational runners," *American Journal of Physical Anthropology*, 81:195–196 (1990); and P. T. Ellison, "Human reproductive ecology in urban environments," in *Human Biology in Urban Environments*, ed. L. Schell and S. Ulijaszek (Cambridge: Cambridge University Press, 1999), pp. 111–135.

184–185 M. M. Fichter and K. M. Pirke, "Hypothalamic-pituitary function in starving healthy subjects," in *The Psychobiology of Anorexia Nervosa*, ed. K. M. Pirke and D. Ploog (Berlin: Springer-Verlag, 1984), pp. 124–135.

185 For weight change and ovarian function, see C. Lager and P. T. Ellison, "Effect of moderate weight loss on ovarian function assessed by salivary progesterone measurements," *American Journal of Human Biology*, 2:303–312 (1990), and Lipson and Ellison, "Comparison of salivary steroid profiles in naturally occurring conception and non-conception cycles.

186 Evolutionary implications of the effect of exercise on ovarian function are discussed by G. R. Bentley, "Hunter-gatherer energetics and fertility: a reassessment of the !Kung San," *Human Ecology*, 13:79–109 (1985), and N. R. Peacock, "Comparative and cross-cultural approaches to the study of human female reproductive failure," in *Primate Life History and Evolution*, ed. C. J. De Rousseau (New York: Wiley-Liss, 1990), pp. 195–220.

186–189 Studies of Lese reproductive ecology are summarized in P. T. Ellison, N. R. Peacock, and and C. Lager, "Ecology and ovarian function among Lese females of the Ituri Forest, Zaire," *American Journal of Physical Anthropology*, 78:519–526 (1989); P. T. Ellison, N. R. Peacock, and C. Lager, "Salivary progesterone and luteal function in two low-fertility populations of northeast Zaire," *Human Biology*, 58:473–483 (1986); R. C. Bailey, M. R. Jenike, P. T. Ellison, G. R. Bentley, A. M. Harrigan, and N. R. Peacock, "The ecology of birth seasonality among agriculturalists in Central Africa," *Journal of Biosocial Science*, 24:393–412 (1992); and G. R. Bentley, A. M. Harrigan, and P. T. Ellison, "Dietary composition and ovarian function among Lese horticulturalist women of the Ituri Forest, Democratic Republic of Congo," *European Journal of Clinical Nutrition*, 52:261–270 (1998).

189–192 For studies of the reproductive ecology of the Tamang, see C. Panter-Brick, D. Lotstein, and P. T. Ellison, "Seasonality of reproductive function and weight loss in rural Nepali women," *Human Reproduction*, 8:684–690 (1993); C. Panter-Brick and P. T. Ellison, "Seasonality of workloads and ovarian function in Nepali women," *Annals of the New York Academy of Sciences*, 709:234–235 (1994); and C. Panter-Brick, "Proximate determinants of birth seasonality and conception failure in Nepal," *Population Studies*, 50:203–220 (1996).

192–193 The study of Polish farmwomen is summarized in G. Jasienska and P. T. Ellison, "Physical work causes suppression of ovarian function in women," *Proceedings of the Royal Society*, Series B, 265:1847–1854 (1998).

194–195 The general phenomenon of human birth seasonality is described in S. Becker, "Seasonal patterns of births and conception throughout the world," in *Temperature and Environmental Effects on the Testis*, ed. A. W. Zorgniotti (New York: Plenum, 1991), pp. 59–72, and D. A. Lam and J. A. Miron, "Global patterns of seasonal variation in human fertility," *Annals of the New York Academy of Sciences*, 709:9–28 (1994). For arguments for and against the importance of seasonal patterns of coital frequency, see J. R. Udry and N. M. Morris, "Seasonality of coitus and seasonality of birth," *Demography*, 4:673–681 (1967), and Bailey et al., "The ecology of birth seasonality among agriculturalists in Central Africa."

195–197 Studies of birth seasonality in Matlab are summarized in J. Stoeckel, A. K. M. A. Chowdhury, "Seasonal variation in births in rural East Pakistan," *Journal of Biosocial Science*, 4:107–116 (1972); S. Becker, A. Chowdhury, and H. Leridon, "Seasonal patterns of reproduction in Matlab, Bangladesh," *Population Studies*, 40:457–472 (1986); and S. Becker, "Understanding seasonality in Bangladesh," *Annals of the New York Academy of Sciences*, 709:370–378 (1994).

197–198 Turkana ecology and birth seasonality are discussed in Michael A. Little and Paul W. Leslie, eds., *Turkana Herders of the Dry Savanna: Ecology and Biobehavioral Response of Nomads to an Uncertain Environment* (Oxford: Oxford University Press, 1999), and P. W. Leslie and P. H. Fry, "Extreme seasonality of births among nomadic Turkana pastoralists," *American Journal of Physical Anthropology*, 79:103–115 (1989).

198–201 Birth seasonality among the Lese is described and discussed in

Bailey et al., "The ecology of birth seasonality among agriculturalists in Central Africa," and M. R. Jenike, R. Bailey, P. T. Ellison, G. R. Bentley, A. M. Harrigan, and N. R. Peacock, "Variation saisonnière de la production alimentaire, statut nutritionnel, fonction ovarienne et fécondité en Afrique centrale," in *L'alimentation en forêt tropical: interactions bioculturelles et perspectives de developpement*, ed. C. M. Hladik, A. Hladik, H. Pagezy, O. F. Linares, and A. Froment (Paris: UNESCO, 1996), pp. 605–623.

201–202 For the importance of pulsatile gonadotropin secretion, see R. M. Boyar, J. Katz, J. W. Finkelstein, S. Kapen, H. Weiner, E. D. Weitzman, and L. Hellman, "Anorexia nervosa: immaturity of the 24-hour luteinizing hormone secretory pattern," *New England Journal of Medicine*, 291:861–865 (1974); R. A. Vigersky, A. E. Anderson, R. H. Thompson, and D. L. Loriaux, "Hypothalamic dysfunction in secondary amenorrhea associated with simple weight loss," *New England Journal of Medicine*, 297:1141–1145 (1977); G. Dixon, P. Eurman, B. Stern, B. Schwartz, and R. W. Rebar, "Hypothalamic function in amenorrheic runners," *Fertility and Sterility*, 42:377–383 (1984); and J. D. Veldhuis, W. S. Evans, L. M. Demers, M. O. Thorner, D. Wakat, and A. D. Rogol, "Altered neuroendocrine regulation of gonadotropin secretion in women distance runners," *Journal of Clinical Endocrinology and Metabolism*, 6:557–563 (1985).

202–203 For evidence of variation in ovarian function without variation in gonadotropin patterns, see K. M. Pirke, U. Schweiger, A. Broocks, B. Spyra, R. J. Tuschl, and R. G. Laessle, "Endocrine studies in female athletes with and without menstrual disturbances," in *The Menstrual Cycle and Its Disorders*, pp. 171–178, and D. C. Cumming, "Menstrual disturbances caused by exercise," in *The Menstrual Cycle and Its Disorders*, pp. 150–160. For the process of extragonadal estrogen formation, see A. Nimrod and K. J. Ryan, "Aromatization of androgens by human abdominal and breast fat tissue," *Journal of Clinical Endocrinology and Metabolism*, 40:367–372 (1975), and P. C. MacDonald, C. D. Edman, D. L. Hemsell, J. C. Porter, and P. K. Siiteri, "Effect of obesity on conversion of plasma androstenedione to estrone in postmenopausal women with and without endometrial cancer," *American Journal of Obstetrics and Gynecology*, 130:448–455 (1978). For variation in estrogen metabolism, see J. Fishman, R. M. Boyar, and L. Hellman, "Influence of body weight on estradiol metabolism in young women," *Journal of Clinical Endocrinology and Metabolism*, 41:989–991 (1975). For the importance of feedback regulation, see P. T. Ellison, "Correlations of basal estrogens with adrenal androgens and relative weight in normal women," *Annals of Human Biology*, 11:327–336 (1984).

203 For cortisol and energetic stress, see A. Loucks, J. Mortola, L. Girton, and S. Yen, "Alterations in the hypothalamic-pituitary-ovarian and the hypothalamic-pituitary-adrenal axes in athletic women," *Journal of Clinical Endocrinology and Metabolism*, 68:402–411 (1989); A. B. Loucks, "Effects of exercise training on the menstrual cycle: existence and mechanisms," *Medicine and Science in Sports and Exercise*, 22:275–80 (1990); and M. Ferin, "Stress and the reproductive cycle," *Journal of Clinical Endocrinology and Metabolism*, 84:1768–1774 (1999).

203–204 For insulin effects, see J. Henriksson, "Influence of exercise on insulin sensitivity," *Journal of Cardiovascular Risk*, 2:303–309 (1995); R. Nahum, K. J. Thong, and S. G. Hillier, "Metabolic regulation of androgen production by human thecal cells in vitro," *Human Reproduction*, 10:75–81 (1995); D. Willis and S. Franks, "Insulin action in human granulosa cells from normal and polycystic ovaries is mediated by the insulin receptor and not the type-I insulin-like growth factor receptor," *Journal of Clinical Endocrinology and Metabolism*, 80:3788–3790 (1990); D. Willis, H. Mason, C. Gilling-Smith, and S. Franks, "Modulation by insulin of follicle-stimulating hormone and luteinizing hormone actions in human granulosa cells of normal and polycystic ovaries," *Journal of Clinical Endocrinology and Metabolism*, 81:302–309 (1996); A. J. Duleba, R. Z. Spaczynski, and D. L. Olive, "Insulin and insulin-like growth factor I stimulate the proliferation of human ovarian theca-interstitial cells," *Fertility and Sterility*, 69:335–340 (1998); and L. Poretsky, N. A. Cataldo, Z. Rosenwaks, and L. C. Giudice, "The insulin-related ovarian regulatory system in health and disease," *Endocrine Reviews*, 20:535–582 (1999).

204–205 For the effects of growth hormone and IGF-I, see E. Y. Adashi, C. E. Resnick, A. Hurwitz, E. Ricciarelli, E. R. Hernandez, C. T. Roberts, D. Leroith, and R. Rosenfeld, "Insulin-like growth factors: the ovarian connection," *Human Reproduction*, 6:1213–1219 (1991); X. C. Jia, J. Kalmin, and A. J. W. Hsueh, "Growth hormone enhances follicle-stimulating hormone-induced differentiation of cultured rat granulosa cells," *Endocrinology*, 118:1401–1409 (1986); L. A. Hutchinson, J. K. Findlay, and A. C. Herrington, "Growth hormone and insulin-like growth factor-I accelerate PMSG-induced differentiation of granulosa cells," *Molecular and Cellular Endocrinology*, 55:61–69 (1998); Y. Yoshimura, Y. Makamura, N. Koyama, M. Iwashita, T. Adachi, and Y. Takeda, "Effects of growth hormone on follicle growth, oocyte maturation, and ovarian steroidogenesis," *Fertility and Sterility*, 59:917–923 (1993); H. D. Mason, H. Martikainen, R. W. Beard, V. Anyaoku, and S. Franks, "Direct gonadotrophic effect of growth hormone on oestradiol production by human granulosa cells in vitro," *Journal of Endocrinology*, 126:R1–4 (1990); J. Tapanainen, H. Martikainen, R. Voutilainen, M. Orava, A. Ruokonen, and L. Ronnberg, "Effect of growth hormone administration on human ovarian function and steroidogenic gene expression in granulosa-luteal cells," *Fertility and Sterility*, 58:726–732 (1992); M. Ando, Y. Yoshimura, M. Iwashita, T. Oda, M. Karube, Y. Ubukata, M. Jinno, and Y. Nakamura, "Direct ovarian effect of growth hormone in the rabbit," *American Journal of Reproductive Immunology*, 31:123–132 (1994); F. I. Sharara and L. K. Nieman, "Identification and cellular localization of growth hormone receptor gene expression in the human ovary," *Journal of Clinical Endocrinology and Metabolism*, 79:670–672 (1994); L. C. Giudice, "Insulin-like growth factors and ovarian follicular development," *Endocrine Reviews*, 13:641–663 (1992); J-H. Olsson, B. Carlsson, and T. Hillensjo, "Effect of insulin-like growth factor-I on deoxyribonucleic acid synthesis in cultured human granulosa cells," *Fertility and Sterility*, 54:1052–1057 (1990); E. L. Yong, D. T. Baird, R. Yates, L. E. Reichert, Jr., and S. G. Hillier, "Hormonal regulation of the growth and steroidogenic function of human granulosa cells," *Journal of Clinical Endocrinology and Metabolism*, 74:842–849 (1992); L. Devoto, P. Kohen, O. Castro, M. Vega, J. L. Troncoso, and E. Charreau, "Multihormonal regulation of progesterone synthesis in cultured human midluteal cells," *Journal of Clinical Endocrinology and Metabolism*, 80:1566–1570 (1995); L. Obassiolu, F. Khan-Dawood, and Y. Dawood, "Insulin-like growth factor-I receptors in human corpora lutea," *Fertility and Sterility*, 37:1235–1240 (1992); E. Hernandez, A. Hurwitz, A. Vera, A. Pellicer, E. Y. Adashi, D. LeRoith, and C. T. Roberts, Jr., "Expression of the genes encoding the insulin-like growth factors and their receptors in the human ovary," *Journal of Clinical Endocrinology and Metabolism*, 74:419–425 (1992); M. L. Hartman, J. D. Veldhuis, and M. O. Thorner, "Normal control of growth hormone secretion," *Hormone Research*, 40:37–47 (1993); J. P. Thissen, J. M. Ketelslegers, and L. E. Underwood, "Nutritional regulation of the insulin-like growth factors," *Endocrine Reviews*, 15:80–95 (1994); and M. L. Vance, M. L. Hartman, and M. O. Thorner, "Growth hormone and nutrition," *Hormone Research*, Supplement 38:85–88 (1992).

205–206 For leptin, see J. S. Flier, "What's in a name? In search of leptin's physiologic role," *Journal of Clinical Endocrinology and Metabolism*, 83:1407–1413 (1998); K. Clément, C. Vaisse, N. Lahlou, S. Cabrol, V. Pelloux, D. Cassuto, M. Gourmelen, C. Dina, J. Chambaz, P. Bougnères, Y. Lebouc, P. Froguel, and B. Guy-gran, "A mutation in the human leptin receptor gene causes obesity and pituitary dysfunction," *Nature*, 392:398–401 (1998); S. B. Heymsfield, A. S. Greenberg, K. Fujioka, R. M. Dixon, R. Kushner, T. Hunt, J. A. Lubina, J. Patane, B. Self, P. Hunt, and M. McCamish, "Recombinant leptin for weight loss in obese and lean adults: a randomized, controlled, dose-escalation trial," *Journal of the American Medical Association*, 282:1568–1575 (1999); M. Rosenbaum, M. Nicolson, J. Hirsch, E. Murphy, F. Chu, and R. L. Leibel, "Effects of weight change on plasma leptin concentrations and energy expenditure," *Journal of Clinical Endocrinology and Metabolism*, 82:3647–3654 (1997); T. A. Warden, R. V. Considine, G. D. Foster, D. A. Anderson, D. B. Sarwer, and J. S. Caro, "Short- and long-term changes in serum leptin in dieting and obese women: effects of caloric restriction and weight loss," *Journal of Clinical Endocrinology and Metabolism*, 83:214–218 (1998); K. K. Miller, M. S. Parulekar, E. Schoenfeld, E. Anderson, J. Hubbard, A. Klibanski, and S. K. Grinspoon, "Decreased leptin levels in normal weight women with hypothalamic amenorrhea: the effects of body composition and nutritional intake," *Journal of Clinical Endocrinology and Metabolism*, 83:2309–2312 (1998); F. F. Chehab, K. Mounzih, R. Lu, and M. E. Lim, "Early onset of reproductive function in normal female mice treated with leptin," *Science*, 275:88–90 (1997); C. C. Cheung, J. E. Thornton, J. L. Kuijper, D. S. Weigle, D. K. Clifton, and R. A. Steiner, "Leptin is a metabolic gate for the onset of puberty in the female rat," *Endocrinology*, 138:855–858 (1997); T. M. Plant and A. R. Durant, "Circulating leptin does not appear to provide a signal for triggering the initiation of puberty in the male rhesus monkey (*Macaca mulatta*)," *Endocrinology*, 138:4505–4508 (1997); D. Apter, "Leptin in puberty," *Clinical Endocrinology*, 47:174–176 (1997); C. Karlsson, K. Lindell, E. Svensson, C. Bergh, P. Lind, H. Bilig, L. M. S. Carlsson, and B. Carlsson, "Expression of functional leptin receptors in the human ovary," *Journal of Clinical Endocrinology and Metabolism*, 82:4144–4148 (1997); J. L. Spicer and C. C. Francisco, "The adipose gene product, leptin: evidence of a direct inhibitory role in ovarian function," *Endocrinology*, 138:3374–3379 (1997); L. Hardie, P. Trayhurn, D. Abramovich, and P. Fowler, "Circulating leptin in women: a longitudinal study in the menstrual cycle and during pregnancy," *Clinical Endocrinology*, 47:101–106

(1997); F. Geisthövel, A. Meysing, and G. Brabant, "C-peptide and insulin, but not C19-steroids, support the predictive value of body mass index on leptin in serum of premenopausal women," *Human Reproduction*, 13:547–553 (1998); X. Casabiell, V. Piñeiro, R. Peino, M. Lage, J. Camiña, R. Gallego, L. G. Vallejo, C. Dieguez, and F. F. Casanueva, "Gender differences in both spontaneous and stimulated leptin secretion by human omental adipose tissue *in vitro:* dexamethasone and estradiol stimulate leptin release in women, but not in men," *Journal of Clinical Endocrinology and Metabolism*, 83:2149–2155 (1998); M. R. Palmert, S. Radovick, and P. A. Boepple, "The impact of reversible gonadal sex steroid suppression on serum leptin concentrations in children with central precocious puberty," *Journal of Clinical Endocrinology and Metabolism*, 83:1091–1096 (1998); N. D. Quinton, R. F. Smith, P. E. Clayton, M. S. Gill, S. Shalet, S. K. Justice, S. A. Simon, S. Walters, M.-C. Postel-vinay, A. I. F. Blakemore, and R. J. M. Ross, "Leptin binding activity changes with age: the link between leptin and puberty," *Journal of Clinical Endocrinology and Metabolism*, 83:2336–2341 (1998); G. Wiesner, M. Vaz, G. Collier, D. Seals, D. Kaye, G. Jennings, G. Lambert, D. Wilkinson, and M. Esler, "Leptin is released from the human brain: influence of adiposity and gender," *Journal of Clinical Endocrinology and Metabolism*, 84:2270–2274 (1999); M. Rosenbaum and R. L. Leibel, "Role of gonadal steroids in the sexual dimorphisms in body composition and circulating concentrations of leptin," *Journal of Clinical Endocrinology and Metabolism*, 84:1784–1789 (1999); J. D. Brannian, Y. Zhao, and M. McElroy, "Leptin inhibits gonadotrophin-stimulated granulosa cell progesterone production by antagonizing insulin action," *Human Reproduction*, 14:1445–1448 (1999); and E. Carmina, M. Ferin, F. Gonzalez, and R. Lobo, "Evidence that insulin and androgens may participate in the regulation of serum leptin levels in women," *Fertility and Sterility*, 72:926–931 (1999).

207 For discussion of pulsatile progesterone patterns and their measurement, see G. R. Merriam and K. Wachter, "Algorithms for the study of episodic hormone secretion," *American Journal of Physiology*, 243:E310-E318 (1982); M. T. O'Rourke and P. T. Ellison, "Salivary measurement of episodic progesterone release," *American Journal of Physical Anthropology*, 81:423–428 (1990); T. M. Delfs, P. Fottrell, S. Klein, O. G. Naether, F. A. Leidenberger, and R. C. Zimmermann, "24-hour profiles of salivary progesterone," *Fertility and Sterility*, 62:960–966 (1994).

210 Jasienska's ideas on energy flux and ovarian function are discussed

in G. Jasienska, *Energy Expenditure and Ovarian Function in Rural Women from Poland* (Ann Arbor, : University Microfilms, 1996), and Jasienska and Ellison, "Physical work causes suppression of ovarian function in women."

211 For discussion of temperature and photoperiod effects on human birth seasonality, see Lam and Mirion, "Global patterns of seasonal variation in human fertility," and Richard J. Levine, "Male factors contributing to the seasonality of human reproduction," *Annals of the New York Academy of Sciences*, 709:29–45 (1994).

212–213 Chronic environmental effects on ovarian function are discussed in P. T. Ellison, "Age and developmental effects on human ovarian function," in *Variability in Human Fertility: A Biological Anthropological Approach*, ed. L. Rosetta and N. C. G. Mascie-Taylor (Cambridge: Cambridge University Press, 1996), pp. 69–90.

The Arc of Life

217 The French study of artificial insemination by donor is presented in Fédération CECOS, D. Schwartz, and M. J. Mayaux, "Female fecundity as a function of age," *New England Journal of Medicine*, 306:404–406 (1982). The accompanying editorial is A. H. DeCherney and G. S. Berkowitz, "Female fecundity and age," *New England Journal of Medicine*, 306:424–426 (1982).

217–218 The initial report of successful oocyte donation is M. V. Sauer, R. J. Paulson, and R. A. Lobo, "A preliminary report on oocyte donation extending reproductive potential to women over 40," *New England Journal of Medicine*, 323:1157–1160 (1990). The *Boston Globe* article is R. Saltus, "When the biological clock pauses," *Boston Globe*, October 30, 1990, pp. 1, 4.

219 The quotation from Louis Henry is from "Some data on natural fertility," *Eugenics Quarterly*, 8:86 (1961).

219–220 For evidence of waning coital frequency with age, see J. R. Udry and N. M. Morris, "Relative contribution of male and female age to the frequency of intercourse," *Social Biology*, 25:128–134 (1978); J. R. Udry, "Changes in the frequency of marital intercourse from panel data," *Archives of Sexual Behavior*, 9:319–325 (1980); and J. R. Udry, F. R. Deven, and S. J. Coleman, "A cross-national comparison of the relative influence of male and female age on the frequency of marital intercourse," *Journal of Biosocial Science*, 14:1–6 (1982).

220 For models of biological and behavioral effects of age on female fertility, see M. Weinstein, J. Wood, M. A. Stoto, and D. D. Green-

field, "Components of age-specific fecundability," *Population Studies*, 44:447–467 (1990); M. Weinstein, J. Wood, and C. Ming-Cheng, "Age patterns of fecundability," in *Biomedical and Demographic Determinants of Human Reproduction*, ed. R. Gray and J. Hobcraft (Oxford: Clarendon Press, 1993), pp. 209–227.

220–221 The evidence from oocyte donation studies is reviewed in P. T. Ellison, "Advances in human reproductive ecology," *Annual Review of Anthropology*, 23:255–275 (1994).

221–222 The concept of adolescent sterility is advanced in M. F. Ashley Montagu, *Adolescent Sterility: A Study in the Comparative Physiology of the Infecundity of the Adolescent Organism in Mammals and Man* (Springfield, Ill.: C. C. Thomas, 1946). Age variation in menstrual patterns is documented in A. E. Treloar, R. E. Boynton, B. G. Behn, and B. W. Brown, "Variation of the human menstrual cycle through reproductive life," *International Journal of Fertility*, 12:77–126 (1967).

222–223 Döring's studies are summarized in G. K. Döring, "The incidence of anovular cycles in women," *Journal of Reproduction and Fertility*, Supplement 6:77–81 (1969).

223 Age variation in salivary progesterone profiles is described in S. F. Lipson and P. T. Ellison, "Normative study of age variation in salivary progesterone profiles," *Journal of Biosocial Science*, 24:233–244 (1992), and S. F. Lipson and P. T. Ellison, "Reference values for luteal progesterone measured by salivary radioimmunoassay," *Fertility and Sterility*, 61:448–454 (1994).

223–224 Apter and Vihko's studies are described in D. Apter, L. Viinkka, and R. Vihko, "Hormonal pattern of adolescent menstrual cycles," *Journal of Clinical Endocrinology and Metabolism*, 47:944–954 (1978); R. Vihko and D. Apter, "Endocrine characteristics of adolescent menstrual cycles: impact of early menarche," *Journal of Steroid Biochemistry*, 20: 231–236 (1984); D. Apter, I. Raisanen, P. Ylostalo, and R. Vihko, "Follicular growth in relation to serum hormonal patterns in adolescents compared with adult menstrual cycles," *Fertility and Sterility*, 47:82–88 (1987); and D. Apter and R. Vihko, "Serum sex hormone-binding globulin and sex steroids in relation to pubertal and postpubertal development of the menstrual cycle," *Progress in Reproductive Biology and Medicine*, 14:58–69 (1990).

224 Age differences in pulsatile progesterone patterns are described in P. T. Ellison, M. T. O'Rourke, R. E. Bledsoe, C. T. Thorne, M. C. Eakin, and S. F. Lipson, "Evidence for two modes of suppression of

luteal phase progesterone in humans," *American Journal of Physical Anthropology*, Supplement 12:70–71 (1991).

224–225 Comparison of age changes in ovarian function in different populations is presented in P. T. Ellison, C. Panter-Brick, S. F. Lipson, and M. T. O'Rourke, "The ecological context of human ovarian function," *Human Reproduction*, 8:2248–2258 (1993); and P. T. Ellison, "Age and developmental effects on human ovarian function," in *Variability in Human Fertility: A Biological Anthropological Approach*, ed. L. Rosetta and N. C. G. Mascie-Taylor (Cambridge: Cambridge University Press, 1996), pp. 69–90.

225–226 The effects of maternal age on birth outcomes are discussed in S. Shapiro, E. R. Schlesinger, and R. E. L. Nesbitt, Jr., *Infant, Perinatal, Maternal, and Childhood Mortality in the United States* (Cambridge, Mass.: Harvard University Press, 1968); L. G. Cooper, N. L. Leland, and G. Alexander, "Effect of maternal age on birth outcomes among young adolescents," *Social Biology*, 42:22–35 (1995); and R. L. Goldenberg and L. V. Klerman, "Adolescent pregnancy—another look," *New England Journal of Medicine*, 332:1161–1162 (1995).

227–228 For a discussion of the diagnosis of LPD, see M. R. McNeely and M. R. Soules, "The diagnosis of luteal phase deficiency: a critical review," *Fertility and Sterility*, 50:1–15 (1988).

228 Age-related changes in salivary estradiol are described in M. T. O'Rourke and P. T. Ellison, "Age and prognosis in premenopausal breast cancer," *Lancet*, 342:60 (1993), and M. T. O'Rourke and P. T. Ellison, "Salivary estradiol levels decrease with age in healthy, regularly-cycling women," *Endocrine Journal*, 1:487–494 (1993).

229 For pulsatile progesterone patterns in older women, see M. T. O'Rourke, S. F. Lipson, and P. T. Ellison, "Ovarian function in the latter half of the reproductive lifespan," *American Journal of Human Biology*, 8:751–760 (1996).

230 The effect of donor and recipient age in oocyte donation is documented in D. Levran, I. Ben-Shlomo, J. Dor, Z. Ben-Rafael, L. Nebel, and S. Mashiach, "Age of endometrium and oocytes: observations on conception and abortion rates in an egg donation model," *Fertility and Sterility*, 56:1091–1094 (1991).

230–231 For age effects on oocyte quality, see E. B. Hook, "Rates of chromosome abnormalities at different maternal ages," *Obstetrics and Gynecology*, 58:282–285 (1981), and F. S. vom Saal and C. E. Finch, "Reproductive senescence: phenomena and mechanisms in mammals and selected vertebrates," in *The Physiology of Reproduction*, ed. E. Knobil, J. D. Neill, L. L. Ewing, G. S. Greenwald, C. L. Markert, and D. W. Pfaff (New York: Raven Press, 1988), pp. 2351–2413. For paternal effects on chromosomal defects, see P. A. Parsons, "Parental age and the offspring," *Quarterly Review of Biology*, 39:258–275 (1964), L. S. Penrose, "Mutation," in *Recent Advances in Human Genetics*, ed. L. S. Penrose and H. L. Brouer (London: Churchill, 1961), pp. 1–18; and L. S. Penrose and H. L. Brouer, "Paternal age and mongolism," *Lancet*, 1:1101 (1962).

232 Evidence for gradual age-related decline in ovarian function is found in J. P. Toner, C. B. Philiput, G. S. Jones, and S. J. Muasher, "Basal follicle-stimulating hormone level is a better predictor of in vitro fertilization performance than age," *Fertility and Sterility*, 55:784–791 (1991), and J. P. Toner and R. T. Scott, "Chronologic versus ovarian age—impact on pregnancy among infertile couples," *Seminars in Reproductive Endocrinology*, 13:1–15 (1995).

233–234 Evidence of follicular depletion at menopause is found in E. Block, "Quantitative morphological investigations of the follicular system in women, variations at different ages," *Acta Anatomica*, 14:108–123 (1952); S. Richardson, V. Senikas, and J. F. Nelson, "Follicular depletion during the menopausal transition: evidence for accelerated loss and ultimate complete exhaustion at menopause," *Journal of Clinical Endocrinology and Metabolism*, 65:1231–1237 (1987); A. Gougeon, R. Echocard, and J. C. Thalabard, "Age-related changes of the population of human ovarian follicles: increase in the disappearance rate of non-growing and early-growing follicles in aging women," *Biology of Reproduction*, 50:653–663 (1994); and G. J. Scheffer, F. J. M. Broekmans, M. Dorland, J. D. F. Habberna, C. W. N. Looman, and E. R. te Velde, "Antral follicle counts by transvaginal ultrasonography are related to age in women with proven natural fertility," *Fertility and Sterility*, 72:845–851 (1999). For the physiology of hot flashes, see D. R. Meldrum, I. V. Tataryn, A. M. Frumar, Y. Erlik, K. H. Lu, and H. L. Judd, "Gonadotropins, estrogens, and adrenal steroids during the menopausal hot flash," *Journal of Clinical Endocrinology and Metabolism*, 50:685–689 (1980).

234–235 Follicular selection is described in G. S. Greenwald and P. F. Terranova, "Follicular selection and its control," in *The Physiology of Reproduction*, pp. 387–445. The functional significance of a dwindling follicular pool is discussed in M. T. O'Rourke, *Human Ovarian Function in Late Reproductive Life* (Ann Arbor: University Microfilms, 1992.)

235 Reproductive senescence in laboratory rodents is discussed in vom

Saal and Finch, "Reproductive senescence: phenomena and mechanisms in mammals and selected vertebrates."

235–236 Evidence of follicular depletion and menopause in other mammals is presented in A. M. Mandl and M. Shelton, "A quantitative study of oöcytes in young and old nulliparous laboratory rats," *Journal of Endocrinology*, 18:444–450 (1959); B. H. Erickson, "Development and senescence of the postnatal bovine ovary," *Journal of Animal Science*, 25:800–805 (1966); S. H. Green and S. Zuckerman, "The number of oöcytes in the mature rhesus monkey (*Macaca mulatta*)," *Journal of Endocrinology*, 7:194–202 (1951); S. H. Green and S. Zuckerman, "Further observations on oöcyte numbers in mature rhesus monkeys (*Macaca mulatta*)," *Journal of Endocrinology*, 10:284–290 (1954); G. D. Hodgen, A. L. Goodman, A. O'Conor, and D. K. Johnson, "Menopause in rhesus monkeys: model for study of disorders in the human climacteric," *American Journal of Obstetrics and Gynecology*, 127:581–584 (1977); T. M. Caro, D. W. Sellen, A. Paris, R. Frank, D. M. Brown, E. Volan, and M. Borgerhoff Mulder, "Termination of reproduction in non-human and human female primates," *International Journal of Primatology*, 16: 205–220 (1995); and M. S. McD. Pavelka and L. M. Fedigan, "Reproductive termination in female Japanese Monkeys: a comparative life history perspective," *American Journal of Physical Anthropology*, 109:455–464 (1999).

236 For evolutionary arguments regarding the recent evolution of menopause, see M. S. McD. Pavelka and L. M. Fedigan, "Menopause: a comparative life history perspective," *Yearbook of Physical Anthropology*, 34:13–38 (1991); K. M. Weiss, "Evolutionary perspectives on human aging," in *Other Ways of Growing Old*, ed. P. Amoss and S. Harrell (Stanford: Stanford University Press, 1981), pp. 25–52; and S. L. Washburn, "Longevity in primates," in *Aging, Biology, and Behavior*, ed. J. March and J. McGaugh (New York: Academic Press, 1981).

236 For the basic evolutionary theory of senescence, see G. C. Williams, "Pleiotropy, natural selection, and the evolution of senescence," *Evolution*, 11:398–411 (1957), and W. D. Hamilton, "The moulding of senescence by natural selection," *Journal of Theoretical Biology*, 12:12–45 (1966).

237–238 For the time-depth of human longevity, see G. Acsadi and J. Nemeskeri, *History of Human Life Span and Mortality* (Budapest: Akadémiai Kiadó, 1970). For the development of the grandmother hypothesis, see R. D. Alexander, "The evolution of social behavior," *Annual Review of Ecology and Systematics*, 5:325–383 (1974); S.

Gaulin, "Sexual dimorphism in the human post-reproductive life-span: possible causes," *Human Evolution*, 9:227–232 (1980); K. Hawkes, J. F. O'Connell, and N. Blurton-Jones, "Hardworking Hadza grandmothers," in *Comparative Socioecology of Mammals and Man*, ed. V. Stardon and R. Foley (London: Basil Blackwell, 1989), pp. 341–366; and K. Hawkes, J. F. O'Connell, N. G. Blurton-Jones, H. Alvarez, and E. L. Charnov, "Grandmothering, menopause, and the evolution of human life histories," *Proceedings of the National Academy of Sciences*, 95:1336–9 (1998). For Hill and Hurtado's critique, see K. Hill and A. M. Hurtado, "The evolution of reproductive senescence and menopause in human females," *Human Nature*, 2:315–350 (1991).

238–239 For evolutionary patterns in oogenesis, see L. L. Franchi, A. M. Mandl, and S. Zuckerman, "The development of the ovary and the process of oögenesis," in *The Ovary*, vol. 1, ed. S. Zuckerman (New York: Academic Press, 1962), pp. 1–88.

241–242 The concept of adaptive peaks was first presented in Sewall Wright, "The roles of mutation, inbreeding, crossbreeding, and selection in evolution," *Proceedings of the Sixth International Congress of Genetics*, 1:356–366 (1932). For menopause in whales, see H. Marsh and T. Kasuya, "Evidence for reproductive senescence in female Cetaceans," *Report of the International Whaling Commission*, Special Issue 8:57–74 (1986).

242–243 Issues of scaling life history are discussed in W. A. Calder III, "Body size, longevity and mortality," *Journal of Theoretical Biology*, 102:135–144 (1983), and W. A. Calder III, *Size, Function, and Life History* (Cambridge, Mass.: Harvard University Press, 1984). Comparisons of human and chimpanzee mortality are presented in Kim Hill, "Life history theory and evolutionary anthropology," *Evolutionary Anthropology*, 2:78–88 (1993).

243–244 For longitudinal studies of ovarian function, see D. Apter and R. Vihko, "Early menarche, a risk factor for breast cancer, indicates early onset of ovulatory cycles," *Journal of Clinical Endocrinology and Metabolism*, 57:82–86 (1983); Vihko and Apter, "Endocrine characteristics of adolescent menstrual cycles: impact of early menarche"; D. Apter, N. J. Bolton, G. L. Hammond, and R. Vihko, "Serum sex hormone-binding globulin during puberty in girls and in different types of adolescent menstrual cycles," *Acta Endocrinologica*, 107:413–419 (1984); S. Venturoli, E. Porcu, R. Fabbri, R. Paradisi, S. Ruggeri, G. Bolelli, L. F. Orsini, D. Gabbi, and C. Flamigni, "Menstrual irregularities in adolescents: hormonal pattern and ovarian morphology," *Hormone Research*, 24:269–279

(1986); S. Venturoli, E. Porcu, R. Fabbri, O. Magrini, R. Paradisi, G. Pallotti, L. Gammi, and C. Famigni, "Postmenarchal evolution of endocrine pattern and ovarian aspects in adolescents with menstrual irregularities." *Fertility and Sterility*, 48:78–85 (1987); and J. Gardner and I. Valadian, "Changes over thirty years in an index of gynaecological health," *Annals of Human Biology*, 10:41–55 (1983).

244 For variation in menarcheal age between populations, see Eveleth and Tanner, *Worldwide Variation in Human Growth*.

244–245 Age variation in ovarian function across populations is discussed in Ellison, "Age and developmental effects on human ovarian function," and P. T. Ellison, "Developmental influences on adult ovarian function," *American Journal of Human Biology*, 8:725–734 (1996).

245 The difficulty of studying age at menopause is discussed in J. W. Wood, *Dynamics of Human Reproduction: Biology, Biometry, Demography* (New York: Aldine de Gruyter, 1994).

The Body Builders

257–258 For the general features of male reproductive senescence, see Frederick S. vom Saal and Caleb E. Finch, "Reproductive senescence: phenomena and mechanisms in mammals and selected vertebrates," in *The Physiology of Reproduction*, ed. Ernst Knobil, Jimmy D. Neill, Larry L. Ewing, Gilbert S. Greenwald, Clement L. Markert, and Donald W. Pfaff (New York: Raven Press, 1988), pp. 2351–2413 .

258–259 For the effects of energetic stress on testicular function, see C. J. Bagatell and W. J. Bremmer, "Sperm counts and reproductive hormones in male marathoners and lean controls," *Fertility and Sterility*, 53:688–692 (1990); N. Fellmann, J. Coudert, J. F. Jarrige, M. Bedu, C. Denis, D. Boucher, and J. R. Lacour, "Effects of endurance training on the androgenic response to exercise in man," *International Journal of Sports Medicine*, 6:215–219 (1985); M. Lehmann, U. Gastmann, K. G. Petersen, N. Bachl, A. Seidel, A. N. Khalaf, S. Fischer, and J. Keul, "Training-overtraining: performance, and hormone levels, after a defined increase in training volume versus intensity in experienced middle- and long-distance runners," *British Journal of Sports Medicine*, 26:233–242 (1992); and A. Klibanski, I. Z. Beitins, T. Badger, R. Little, and J. W. McArthur, "Reproductive function during fasting in men." *Journal of Clinical Endocrinology and Metabolism*, 53:258–263 (1981).

259 On sperm counts and male fecundity, see Emmet J. Lamb and Sean

Bennett, "Epidemiologic studies of male factor in infertility," *Annals of the New York Academy of Sciences*, 709:165–178 (1994).

260 Population variation in male testosterone is discussed in P. T. Ellison, S. F. Lipson, R. G. Bribiescas, G. R. Bentley, B. C. Campbell, and C. Panter-Brick, "Inter- and intra-population variation in the pattern of male testosterone by age," *American Journal of Physical Anthropology*, Supplement 26:80 (1998).

261 The linkages between endocrinology and reproductive behavior are reviewed in David Crews, ed., *Psychobiology of Reproductive Behavior* (Englewood Cliffs, N.J.: Prentice-Hall, 1987).

262 For testosterone and copulation rates in rats, see Alan I. Leshner, *An Introduction to Behavioral Endocrinology* (New York: Oxford University Press, 1978). For the treatment of human paraphiliacs, see A. Rösler and E. Witztum, "Treatment of men with paraphilia with a long-acting analogue of gonadotropin-releasing hormone," *New England Journal of Medicine*, 338:416–422 (1998).

262–263 The male contraceptive studies are summarized in R. S. Swerdloff, C. J. Bagatel, C. Wang, B. D. Anawalt, N. Berman, B. Steiner, and W. J. Bremner, "Suppression of spermatogenesis in man induced by Nal-Glu gonadotropin releasing hormone antagonist and testosterone enanthate (TE) is maintained by TE alone," *Journal of Clinical Endocrinology and Metabolism*, 83(10):3527–3533 (1998); C. J. Bagatel, A. M. Matsumoto, R. B. Christensen, J. E. Rivier, and W. J. Bremner, "Comparison of a gonadotropin releasing-hormone antagonist plus testosterone (T) versus T alone as potential male contraceptive regimens," *Journal of Clinical Endocrinology and Metabolism*, 77:427–432 (1993); L. Tom, S. Bhasin, W. Salameh, B. Steiner, M. Peterson, R. Z. Sokol, J. Rivier, W. Vale, and R. S. Swerdloff, "Induction of azoospermia in normal men with combined Nal-Glu gonadotropin-releasing hormone antagonist and testosterone enanthate," *Journal of Clinical Endocrinology and Metabolism*, 75:476–83 (1992).

263 For androgens and libido in human females, see W. Arlt, F. Callies, J. C. van Vlijmen, I. Koehler, M. Reinke, M. Bidlingmaier, D. Huebler, M. Oettel, M. Ernst, H. M. Schulte, and B. Allolio, "Dehydroepiandrosterone replacement in women with adrenal insufficiency," *New England Journal of Medicine*, 341:1013–1020 (1999); S. Davis, "Androgen replacement in women: a commentary," *Journal of Clinical Endocrinology and Metabolism*, 84:1886–1891 (1999); and B. N. Sherwin, M. M. Gelfand, and W. Brender, "Androgen enhances sexual motivation in females: a prospective,

crossover study of sex steroid administration in surgical menopause," *Psychosomatic Medicine*, 47:339–351 (1997).

264 The anonymous account appeared in *Nature*, 226:869 (1970).

264–265 For testosterone and aggression in human males, see S. Bhasin, C. J. Bagatell, W. J. Bremner, S. R. Plymate, J. L. Tenove, S. G. Korenman, and E. Nieschlag, "Issues in testosterone replacement in older men," *Journal of Clinical Endocrinology and Metabolism*, 83:3435–3448 (1998), and J. M. Dabbs, "Salivary testosterone measurements in behavioral studies," *Annals of the New York Academy of Sciences*, 694:177–183 (1993).

265 Dabbs's study of testosterone and occupation is presented in J. M. Dabbs, Jr., D. de La Rue, and P. M. Williams, "Testosterone and occupational choice: actors, ministers, and other men," *Journal of Personality and Social Psychology*, 59:1261–1265 (1990).

266 For dominance and testosterone in primates, see R. M. Rose, T. P. Gordon, and I. S. Bernstein, "Plasma testosterone, dominance rank, and aggressive behavior," *Nature*, 231:366–368 (1971); A. Dixson and J. Herbert, "Testosterone, aggressive behavior, and dominance rank in captive adult male talapoin monkeys (*Miopithecus talapoin*), *Physiology of Behavior*, 18:539–543 (1977); M. S. Golub, E. N. Sassenrath, and G. P. Goo, "Plasma cortisol levels and dominance in peer groups of rhesus monkey weanlings," *Hormones and Behavior*, 12:50–59 (1979); R. M. Rose, I. S. Bernstein, and T. P. Gordon, "Consequences of social conflict on plasma testosterone levels in rhesus monkeys," *Psychosomatic Medicine*, 37:50–61 (1975); I. S. Bernstein, R. M. Rose, and T. P. Gordon, "Behavioral and environmental effects influencing primate testosterone levels," *Journal of Human Evolution*, 3:517–525 (1974); R. M. Rose, I. S. Bernstein, and T. P. Gordon, "Plasma testosterone levels in the male rhesus: effects of sexual and social stimuli," *Science*, 178:643–645 (1972); and M. Muller and R. W. Wrangham, "The reproductive ecology of male hominoids," in *Reproductive Ecology and Human Evolution*, ed. P. T. Ellison (New York: Aldine de Gruyter, forthcoming).

266–268 The first report of competitive outcome effects on human male testosterone is in M. Elias, "Serum cortisol, testosterone, and testosterone-binding globulin responses to competitive fighting in human males," *Aggressive Behavior*, 7:215–224 (1981). Subsequent studies are reviewed in B. A. Gladue, M. Boechler, and D. D. McCaul, "Hormonal response to competition in human males," *Aggressive Behavior*, 15:409–422 (1989); A. Booth, G. Shelley, A. Mazur, G.

Tharp, and R. Kittok, "Testosterone, and winning and losing in human competition," *Hormones and Behavior*, 23:555–571 (1989); A. Mazur, A. Booth, and J. M. Dabbs, Jr., "Testosterone and chess competition," *Social Psychology Quarterly*, 55:70–77 (1992); and Michael W. Rabow, "The agony of defeat: the pulsatile release and response of testosterone to aggressive athletics," B.A. Honors Thesis, Department of Anthropology, Harvard University, 1987.

268–269 An integrated view of human male testosterone and behavioral dynamics is provided in A. Mazur and A. Booth, "Testosterone and dominance in men," *Behavioral and Brain Sciences*, 21:353–397 (1998).

270 For testosterone and male muscle mass, see J. M. Round, D. A. Jones, J. W. Honour, and A. M. Nevill, "Hormonal factors in the development of differences in strength between boys and girls during adolescence: a longitudinal study," *Annals of Human Biology*, 26:49–62 (1999); N. Mauras, V. Hayes, S. Welch, A. Rini, K. Helgeson, M. Dokler, J. D. Veldhuis, and R. J. Urba, "Testosterone deficiency in young men: marked alterations in whole body protein kinetics, strength, and adiposity," *Journal of Clinical Endocrinology and Metabolism*, 83:1886–1892 (1998); and P. J. Snyder, H. Peacey, P. Hannoush, J. A. Berlin, L. Loh, D. A. Lenrow, J. H. Holmes, A. Dlewati, J. Santanna, C. J. Rosen, and B. L. Strom, "Effect of testosterone treatment on body composition and muscle strength in men over 65 years of age," *Journal of Clinical Endocrinology and Metabolism*, 84:2647–2653 (1999).

271 For the metabolic cost of maintaining muscle, see L. C. Aiello, "Human body size and energy," in *The Cambridge Encyclopedia of Human Evolution*, ed. S. Jones, R. Martin, and D. Pilbeam (Cambridge: Cambridge University Press, 1992), pp. 44–45.

271–272 For interactions between testosterone and exercise in determining muscle mass, see S. Bhasin, T. W. Storer, N. Berman, C. Callegari, B. Clevenger, J. Phillips, T. J. Bunnell, R. Tricker, A. Shirazi, and R. Casaburi, "The effects of supraphysiologic doses of testosterone on muscle size and strength in normal men," *New England Journal of Medicine*, 335:1–7 (1996).

273 For evolutionary perspectives on sexual dimorphism in primates, see R. D. Martin, "Sexual dimorphism and the evolution of higher primates," *Nature*, 287:273–275 (1980), and S. J. Gaulin and L. D. Sailer, "Sexual dimorphism in weight among the primates: the relative impact of allometry and sexual selection," *International Journal of Primatology*, 5:515–535 (1984).

274 For sexual rivalry and homicide in humans, see M. Daly and M. Wilson, *Homicide* (New York: Aldine de Gruyter, 1988).

274–276 For evidence of population variation in male testosterone and its correlates, see P. T. Ellison, S. Lipson, and M. Meredith, "Salivary testosterone levels in males from the Ituri Forest, Zaire," *American Journal of Human Biology*, 1:21–24 (1989); K. H. Christiansen, "Serum and saliva sex hormone levels in !Kung San men," *American Journal of Physical Anthropology*, 86:37–44 (1991); C. M. Beall, C. M. Worthman, K. P. Strohl, G. M. Brittenham, and M. Barragan, "Salivary testosterone concentration of Aymara men native to 3600 m.," *Annals of Human Biology*, 19:67–78 (1992); G. R. Bentley, A. M. Harrigan, B. Campbell, and P. T. Ellison, "Seasonal effects on salivary testosterone levels among Lese males of the Ituri Forest, Zaïre," *American Journal of Human Biology*, 5:711–717 (1993); and P. T. Ellison and C. Panter-Brick, "Salivary testosterone levels among Tamang and Kami males of central Nepal," *Human Biology*, 68:955–965 (1996). For a functional interpretation of the variation, see R. G. Bribiescas, "Salivary testosterone levels among Aché hunter/gatherer men and a functional interpretation of population variation in testosterone among adult males," *Human Nature*, 7:163–188 (1996).

277 For age and male potency, see vom Saal and Finch, "Reproductive senescence."

277–278 For age and male testosterone, see N. W. Shock, R. C. Greulich, R. Andres, D. Arenberg, P. T. Costa, E. G. Lakatta, and J. D. Tobin, *Normal Human Aging: The Baltimore Longitudinal Study*, NIH publication no. 84–2450 (Bethesda: National Institutes of Health, 1984); B. C. Campbell, D. J. Colfax, and P. T. Ellison, "Age-related changes in pulsatile release of salivary testosterone among adult males," *American Journal of Physical Anthropology*, Supplement 12:57 (1991); and Ellison et al., "Inter- and intra-population variation in the pattern of male testosterone by age."

The Journey and the Procession

283–286 For general overviews of human evolution, see B. A. Wood, "Origin and evolution of the genus *Homo*," *Nature*, 355:785–790 (1992); B. A. Wood, "Evolution of australopithecines," in *The Cambridge Encyclopedia of Human Evolution*, ed. S. Jones, R. Martin, and D. Pilbeam (Cambridge: Cambridge University Press, 1992), pp. 231–240; C. B. Stringer, "The evolution of early humans," in Jones et al., *The Cambridge Encyclopedia of Human Evolution*, pp. 241–251; and Robert Boyd and Joan B. Silk, *How Humans Evolved* (New York: Norton, 1997). For genetic relationships among humans and apes, see M. Ruvolo, "Molecular phylogeny of the hominoids: inferences from multiple independent DNA sequence data sets," *Molecular Biology and Evolution*, 14:248–265 (1997). For the debate on modern human origins, see G. Bräuer and F. H. Smith, *Replacement or Continuity? Controversies in the Evolution of Homo sapiens* (Rotterdam: Balkema, 1990).

286–287 For recent evidence regarding chimpanzee cultural complexity, see R. W. Wrangham, W. C. McGrew, Frans De Waal, and Paul G. Heltne, eds., *Chimpanzee Cultures* (Cambridge, Mass.: Harvard University Press, 1996).

287–288 Ailello and Wheeler's hypothesis is presented in L. C. Aiello and P. Wheeler, "The expensive tissue hypothesis," *Current Anthropology*, 36:199–211 (1995).

288 For speculation on the importance of meat eating to brain evolution, see K. Milton, "A hypothesis to explain the role of meat-eating in human evolution," *Evolutionary Anthropology*, 8:11–21 (1999), and W. R. Leonard and M. L. Robertson, "Evolutionary perspectives on human nutrition: the influence of brain and body size on diet and metabolism," *American Journal of Human Biology*, 6:77–88 (1994). For the importance of cooking in human evolution, see R. W. Wrangham, J. H. Jones, G. Laden, D. Pilbeam, and N. L. Conklin-Brittain, "The raw and the cooked: cooking and the ecology of human origins," *Current Anthropology*, 40:567–594 (1999).

290 For orangutan ecology, see C. D. Knott, *Reproductive, Physiological, and Behavioral Responses of Orangutans in Borneo to Fluctuations in Food Availability* (Ann Arbor: University Microfilms, 1999); C. D. Knott, "Orangutan behavior and ecology," in *The Nonhuman Primates*, ed. P. Dolhinow and A. Fuentes (Mountain View, Calif.: Mayfield, 1998), pp. 50–57; C. D. Knott, "Changes in orangutan diet, caloric intake, and ketones in response to fluctuating fruit availability," *International Journal of Primatology*, 19:1061–1079 (1998); and C. Knott, "Ape models of female reproductive ecology," in *Reproductive Ecology and Human Evolution*, ed. P. T. Ellison (New York: Aldine de Gruyter, forthcoming).

ACKNOWLEDGMENTS

The world of ideas is exhilarating and rewarding precisely because it is, at its core, a very social world. Without implying that they should share any blame for the errors of my thinking, I would like to thank a number of people for their influences, direct and indirect, on this book.

My teachers and mentors over the years provided inspirational examples of intellectual courage and honesty, especially Irven Devore, Jonathan Friedlaender, Sarah Blaffer Hrdy, Francis Johnston, Nathan Keyfitz, Melvin Konner, James Tanner, Roberta Todd, Eric Trinkaus, Robert Trivers, Alan Walker, and John Whiting. My faculty colleagues in biological anthropology at Harvard have created a wonderful intellectual environment for the formation and testing of ideas without sacrificing a spirit of camaraderie and mutual support. For that, as well as their friendship, I thank John Barry, Nancylou Conklin-Brittain, Irven Devore, Marc Hauser, Cheryl Knott, Mark Leighton, Susan Lipson, Frank Marlowe, David Pilbeam, Maryellen Ruvolo, and Richard Wrangham.

In the laboratory I have benefited from the talented collaboration of my friend and colleague Susan Lipson, as well as the skill and assistance of Marion Eakin, Seema Goel, Grazyna Jasienska, Catherine Lager, Debra Lotsein, Mary O'Rourke, and Sara Sukalich. As a teacher and supervisor I have learned a tremendous amount from my students, teaching fellows, and postdoctoral fellows, including Gillian Bentley, Michael Billig, Richard Bribiescas, Ben Campbell, Judith Flynn, Nancy Friedlander, Alisa Harrigan, Grazyna Jasienska, Cheryl Knott, Anne McGinnis, Matthew McIntyre, Mary O'Rourke, Diana Sherry, and Claudia Valeggia. I owe a special debt to my colleagues and collaborators in the field, especially Robert Bailey, Gillian Bentley, Isabel Bradburn, John Fisher, Alisa Harrigan, Grazyna Jasienska, Mark Jenike, Catherine Panter-Brick, Nadine Peacock, Helen Strickland, Inger Thune, Claudia Valeggia, and Virginia Vitzthum. Other colleagues, friends, and interlocutors have assisted me in many ways through example, discussion, argument, and helpful comments, including Dan Apter, Neville Bruce, Kenneth Campbell, Andrew

Cohn, James Dabbs, Mark Flinn, Kim Hill, Darryl Holman, Sarah Blaffer Hrdy, Hillard Kaplan, Jane Lancaster, William Lasley, Paul Leslie, Nick Mascie-Taylor, Lylianne Rosetta, Robert Sapolsky, Susan Sheideler, Beverly Strassmann, Stanley Ulijaszek, James Woods, and Carol Worthman.

I owe the most profound thanks to the many individuals around the world who have participated in our studies. I am particularly grateful to my friends in the Ituri Forest, whose lives contributed the stories framing the narrative of this book. Their names have been changed in the text according to convention, but their individual characters, voices, and spirits are indelibly inscribed in my heart.

The Guggenheim Foundation provided the fellowship that allowed this book to be written. Major funding for my research has come from the National Science Foundation, the National Institutes of Health, the Department of Defense, the Wenner-Gren Foundation, the Nestlé Foundation, and Smith Kline Beacham. Mary O'Rourke provided essential help with the index. Finally, this book would not have been completed without the faith and assistance of my editors at Harvard University Press, Michael Fisher and Nancy Clemente.

My wife, Pippi, who first introduced me to anthropology, has shared in every aspect of this book. From campfire discussions in the Ituri to critical reading of innumerable drafts, her insights, wisdom, and encouragement have been indispensable. My sons, Sam and Silas, also provided love, support, art, and music, as well as intellectual, athletic, and spiritual challenges all along the way. If all the work represented here were swept away I would still know, because of them, that love and family are what it's all about.

LIST OF ILLUSTRATIONS

INDEX